Wind Energy Storage and Conversion

Scrivener Publishing
100 Cummings Center, Suite 541J
Beverly, MA 01915-6106

Publishers at Scrivener
Martin Scrivener (martin@scrivenerpublishing.com)
Phillip Carmical (pcarmical@scrivenerpublishing.com)

Wind Energy Storage and Conversion

From Basics to Utilities

Edited by

Inamuddin

Tariq Altalhi

and

Mohammad Luqman

Scrivener Publishing

WILEY

This edition first published 2024 by John Wiley & Sons, Inc., 111 River Street, Hoboken, NJ 07030, USA and Scrivener Publishing LLC, 100 Cummings Center, Suite 541J, Beverly, MA 01915, USA

For more information about Scrivener publications please visit www.scrivenerpublishing.com.

Wiley Global Headquarters
111 River Street, Hoboken, NJ 07030, USA

For details of our global editorial offices, customer services, and more information about Wiley products visit us at www.wiley.com.

Library of Congress Cataloging-in-Publication Data

ISBN 978-1-394-20433-5

Front cover images supplied by Adobe Firefly
Cover design by Russell Richardson

Set in size of 11pt and Minion Pro by Manila Typesetting Company, Makati, Philippines

Printed in the USA

10 9 8 7 6 5 4 3 2 1

Contents

Preface **xi**

1 Wind Energy: From Past to Present Technology **1**
Madhur Babu Singh, Pallavi Jain and Prashant Singh
1.1 Introduction 2
1.2 Historical Background 3
1.3 Use of Wind Energy in Specific Countries 5
1.4 Wind Technology 5
1.4.1 Wind Energy Conversion System (WECS) 6
1.4.2 Electric Generator 7
1.4.3 Evolution of Power Electronics 8
1.4.4 Energy Storage Technology 8
1.5 Horizontal-Axis Wind Turbines (HAWTs) 9
1.5.1 History 9
1.5.2 Design 9
1.5.3 Components 10
1.5.4 Working Principle 10
1.5.5 Applications 10
1.6 Vertical Axis Wind Turbine (VAWT) 11
1.6.1 Working Principle 11
1.7 Current Technologies in Wind Power Generation 12
1.7.1 Buoyant Airborne Turbine (BAT) 12
1.7.2 Offshore Floating Wind Technology 12
1.8 Advantages 13
1.9 Disadvantages of Wind Energy 14
1.10 Conclusion 14
References 14

2 Environmental Consequences of Wind Energy Technologies **17**
Adarsh Kumar Arya and Ashish Kapoor
2.1 Introduction 17

2.2 Impact of Wind Energy on the Environment 20
2.3 Key Environmental White Paper Issues Related to Wind Power 21
2.4 Individual Effects on Population Impacts 23
2.5 Comprehending the Overall Effects of Wind Power on Wildlife 24
2.6 Considerations for the Environment when Making Choices 25
2.7 Wind Power and Risk Management 26
2.8 Concerns About Using Wind Energy 27
2.9 Conclusion 29
References 30

3 Important Issues and Future Opportunities for Huge Wind Turbines 33
Anjali Tripathi, Vinay Kumar Pandey, Shivangi Srivastava, Rashi Mishra and Ashish
3.1 Introduction 34
3.1.1 Visual Impact 34
3.1.2 Noise 36
3.1.3 Wildlife 37
3.1.4 Intermittent Energy Generation 39
3.2 Worldwide Wind Energy Forecast 41
3.2.1 Canada 41
3.2.2 Russia 42
3.2.3 India 42
3.2.4 United States of America 43
3.2.5 China 44
3.2.6 Germany 45
3.3 Increased Wind Penetrating Techniques 45
3.3.1 Energy Storage Systems 45
3.3.2 Advanced Forecasting Tools 46
3.3.3 Bucket Foundation 46
3.3.4 Advantages of Bucket Foundation 46
3.3.5 Limitations of Bucket Foundation 47
3.3.6 Monopile Foundation 47
3.3.7 Jacket Foundation 47
3.3.8 Floating Foundation 48
3.3.9 Tripod Foundation 48
3.4 India's Perspective for Wind Energy 49
3.4.1 Intermittency and Variability 51
3.4.2 Land Acquisition 51

3.4.3 Transmission Constraints 51
3.4.4 Limited Wind Resource Data 51
3.4.5 Financing Constraints 52
3.4.6 Environmental and Social Impacts 52
3.4.7 Policy and Regulatory Uncertainty 52
3.5 Progress of Technology 52
3.5.1 Larger and More Efficient Turbines 53
3.5.2 Advancements in Turbine Design 53
3.5.3 Improvements in Manufacturing and Installation 56
3.6 Conclusion 56
References 57

4 Wind Hybrid Power Technologies 63
Momina, Haq Nawaz Bhatti and Amina Khan
4.1 Introduction 63
4.2 Types of Hybrid Power Systems 64
4.3 Wind Hybrid Power Technologies 65
4.3.1 Wind Diesel Hybrid Power Technology 65
4.3.2 Wind Solar Hybrid Power Technology (WSHPT) 70
4.3.3 Wind Hydrogen Hybrid Power Technology (WHHPT) 72
4.3.4 Wind-Hydro Hybrid Power Technology (WHHPT) 75
4.3.5 Wind-Photovoltaic (PV) Hybrid Power Technology 77
4.4 Summary 79
References 79

5 Theories Based on Technological Advances for Wind Energy 83
Muhammad Azam, Laiba Zafar, Haq Nawaz Bhatti and Amina Khan
5.1 Introduction 84
5.2 Theoretical Background 87
5.2.1 Basic Principles of Wind Energy Conversion 87
5.2.2 Aerodynamics of Wind Turbines 88
5.2.3 Control Systems for Wind Turbines 89
5.3 Theories Based on Technological Advances 91
5.3.1 Wind Turbine Design Theory 92
5.3.1.1 Rotor Blade Design Theory 93
5.3.1.2 Aerodynamic Design Theory 94
5.3.2 Power Control Theory 95
5.3.2.1 Maximum Power Point Tracking Theory 96
5.3.2.2 Load Control Theory 98
5.3.3 Wind Farm Layout Theory 99

5.3.3.1 Turbine Placement Theory 100
5.3.3.2 Wake Effect Theory 101
5.3.4 Grid Integration Theory 103
5.3.4.1 Power Quality Theory 104
5.3.4.2 Stability Theory 106
5.4 Advancements in Wind Energy Technologies 107
5.5 Future Research Directions 108
5.6 Conclusion 110
References 111

6 Wind Energy Hybrid Power Generation System with Hydrogen Storage 115
Mehmet Bugdayci and Mesut Yilmazoglu
6.1 Introduction 116
6.2 Hydrogen Storage Systems 120
6.2.1 Solid-State Hydrogen Storage in Materials 121
6.3 Wind Energy Systems 123
6.4 Wind Energy Hybrid Power Generation System with Hydrogen Storage 125
6.4.1 Design and Optimization of a Wind Energy Hybrid Power Generation System with Hydrogen Storage 126
6.5 Conclusion 127
References 128

7 Technologies Based on Reusable Wind Turbine Blades 133
Abhinay Thakur and Ashish Kumar
7.1 Introduction 134
7.2 Wind Power Generation and the Importance of Wind Turbine Blades 137
7.2.1 Global Demand for Clean and Sustainable Energy 137
7.2.2 Role of Wind Turbines in Wind Power Generation 139
7.2.3 Impact of Wind Turbine Blades on Performance and Viability 141
7.3 Conventional Wind Turbine Blade Materials and Limitations 142
7.3.1 Overview of Conventional Blade Materials 142
7.3.2 Limitations in Terms of Recyclability and Environmental Impact 145
7.4 Advancements in Materials Engineering for Reusable Wind Turbine Blades 147
7.4.1 Composite Materials in Blade Design 147
7.4.2 Bio-Based Resins for Sustainable Blades 150

7.4.3 Additive Manufacturing Techniques for Blade Production 153
7.5 Challenges in Implementing Reusable Blade Technologies 155
7.5.1 Structural Integrity of Reusable Blades 155
7.5.2 Fatigue Resistance and Durability 156
7.5.3 Manufacturing Scalability and Cost-Effectiveness 158
7.6 Implications of Reusable Wind Turbine Blades 160
7.6.1 Cost Reduction and Enhanced Energy Production 160
7.6.2 Environmental Benefits and Reduction of Carbon Emissions 162
7.6.3 Policy Frameworks and Industry Collaboration 164
7.7 Testing, Modeling, and Simulation for Reliable Reusable Blade Designs 165
7.7.1 Importance of Rigorous Testing 165
7.7.2 Modeling and Simulation Techniques for Design Optimization 167
7.8 Future Prospects and Research Directions 168
7.8.1 Interdisciplinary Approaches for Sustainable Innovation 168
7.8.2 Collaboration Among Researchers, Engineers, and Stakeholders 170
7.8.3 Potential Directions for Future Research 173
7.9 Conclusion 175
References 177

8 Wind Turbine Assessment: A Step-by-Step Approach 185
Figen Balo and Lutfu S. Sua
8.1 Introduction 186
8.2 Analytic Hierarchy Strategy 191
8.3 Results and Discussion 194
8.4 Conclusions 197
References 201

9 Effect of Aerodynamics on Wind Turbine Design 207
Mahadi Hasan Masud, Md. Forhad Hossain Hemal, Mim Mashrur Ahmed, Md. Fyruz Ibna Alam Taki, Md. Hasibul Hasan Himel, Anan Ashrabi Ananno and Peter Dabnichki
9.1 Introduction 208
9.2 Air Properties Affecting Wind Turbines 210
9.3 Classical Blade Element Momentum Theory 212
9.4 Aerodynamic Performance Testing 215

9.4.1 Wind Tunnel Testing and Field Testing 215
9.4.2 Performance Testing of a Counter-Rotating Wind Turbine System 217
9.5 Effect of Aerodynamics on Wind Turbine Design Parameters 220
9.5.1 Solidity 223
9.5.2 Number of Blades 224
9.5.3 Different Ratios 226
9.5.3.1 Chord/Radius Ratio (c/R) 226
9.5.3.2 Height-to-Radius Ratio (H/R) 227
9.5.3.3 Blade Aspect Ratio (H/c) 229
9.5.4 Pitch 230
9.5.5 Strut Connection Point 232
9.5.6 Blade Reynolds Number (Re) 233
9.5.7 Strut Effects 234
9.5.8 Strut Arrangement 236
9.6 Wind Turbine Loads 238
9.7 Conclusions 239
References 240

Index 247

Preface

With the depletion of fossil fuel-based energy resources, the development of alternative sources of energy is becoming extremely crucial. Meanwhile, the planet is on the brink of an energy disaster due to the rapidly rising global need for energy. Additionally, the widespread use of fossil fuel-based energy resources aggravates global warming and harms the environment. However, there are reliable and eco-friendly substitutes to fossil fuels, like wind and many other sustainable energies. Considering its low operational costs and easy accessibility, the wind is among the most cost-effective and efficient renewable energies. With the increased use of wind energy, the need for storage has become critical. In addition to various storage procedures, fuel cells and batteries are two primary sources of compensation for renewable energy (RE) systems. The wind technological system is on the cusp of development, but numerous improvements are required to make this technology cost-efficient, overall.

In this book, various energy storage and conversion methods for wind power applications are explored. By going through this book, one can learn more about excellent illustrations of the usefulness of adopting renewable energies, particularly in light of the widespread use of wind-based devices and the in-depth presentation of several developments in wind technological systems concerning applications and operational approaches. Additionally, this work covers the costs associated with the electrical output in wind-powered power plants as well as the financial and environmental plans that describe the installation of wind technological systems. For those interested in this field, this publication is a great resource for academics, researchers, environmentalists, and professionals.

Chapter 1 provides a concise scientific overview of wind energy's historical development, technological advancements, and significance as a major renewable energy source. It highlights offshore wind farms, improved efficiency, and reduced greenhouse gas emissions, making wind energy vital for a sustainable future. It also covers ongoing research for enhanced turbine designs.

Chapter 2 analyzes the direct and indirect environmental consequences associated with wind-energy technologies. It examines their potential impacts on living beings and the environment. These concerns may pose challenges to the progress of wind energy projects, underscoring the importance of thoughtful consideration and the implementation of effective mitigation strategies.

Chapter 3 discuss the visual impact of wind turbines and also their impact environment as noise pollution and wild life. The international wind energy generation and world energy forecast is discussed in details.

Chapter 4 discusses the types of wind hybrid power technologies. It focuses on the components and role of each hybrid system. The advantages and disadvantages of each hybrid system are also given. Careful integration of these systems can lessen their shortcomings and generate maximum electric power.

Chapter 5 extensively reviews wind energy theories driven by technological advancements. It starts with the fundamental principles and aerodynamics of wind turbines, moving to practical applications, control systems, and design aspects. It further explores advancements such as innovative materials, turbine designs, enhanced control systems, and smart grid integration, reinforcing wind energy's efficiency and competitiveness.

Chapter 6 discusses the wind-energy hybrid power generation system with hydrogen storage. Additionally, it details hydrogen storage systems and wind energy systems. It concludes with the design and optimization of a wind-energy hybrid power generation system with hydrogen storage.

Chapter 7 investigates the growing global demand for clean and sustainable energy sources and how they have propelled the rapid development of wind power generation. Wind turbines, as key components of this industry, require efficient and cost-effective solutions to maximize their potential. One critical aspect is the design and manufacture of wind turbine blades, which significantly impact the overall performance and economic viability of wind farms. This chapter also explores emerging technologies centered on reusable wind turbine blades, highlighting their advancements, challenges, and implications. The chapter provides an overview of conventional wind turbine blade materials and their limitations in terms of recyclability and environmental impact. It then delves into recent research and innovations in materials engineering, such as composite materials, bio-based resins, and additive manufacturing techniques. Furthermore, the chapter examines the challenges faced in implementing reusable blade technologies, including structural integrity, fatigue resistance, and manufacturing scalability.

Chapter 8 details how to balance and choose the best alternative for the criteria after calculating significant aspects in the evaluation of a wind turbine technology following the principles of sustainable development for analytical hierarchy process (AHP). Based on the findings, the most optimal alternative is chosen.

Chapter 9 discusses different aerodynamic properties like solidity, blade number, pitch, pitch angle, strut effect, and Reynolds number as they directly affect turbine performance. The optimum values of all these parameters for improved performance are also summarized.

Key features:

- Provides a broad overview of wind energy technology
- Describes prospects for energy storage and conversion systems
- Analyzes economic and environmental plans for the development of wind energy systems

1

Wind Energy: From Past to Present Technology

Madhur Babu Singh[1], **Pallavi Jain**[1*] **and Prashant Singh**[2†]

[1]*Department of Chemistry, SRM Institute of Science and Technology, Delhi-NCR Campus, Modinagar, India*

[2]*Department of Chemistry, Atma Ram Sanatan Dharma College, University of Delhi, New Delhi, India*

Abstract

Humans have used wind energy for centuries. In the 7th century, ancient Persia used it to grind grain and pump water. The concept of using wind to produce electricity was initially investigated in the late 19th century, and in recent years, wind energy has emerged as a major player in the renewable energy market, with numerous nations investing in wind power projects. With numerous turbines positioned far from the shore, where the winds are steady and strong, offshore wind farms have grown in popularity. With the potential to produce electricity even in the absence of wind, wind energy has become more reliable and cost-effective thanks to technological improvements. This renewable energy source is crucial in the shift to a more sustainable future since it lowers greenhouse gas emissions while meeting the rising demand for electricity. As the world continues to move towards sustainable sources, wind energy is expected to offer a larger portion of energy needs. Scientists are researching novel wind turbine designs that might increase their output and efficiency. Wind energy is a potential renewable energy technology with a lengthy history and a bright future.

Keywords: Wind energy, turbine, environment, sustainable development

**Corresponding author*: palli24@gmail.com

†*Corresponding author*: psingh@arsd.du.ac.in

Inamuddin, Tariq Altalhi and Mohammad Luqman (eds.) Wind Energy Storage and Conversion: From Basics to Utilities, (1–16) © 2024 Scrivener Publishing LLC

1.1 Introduction

Energy is a crucial aspect of our sustainability, playing a significant role in promoting human civilization and shaping our lives. The social and economic progress of modern society relies heavily on a steady and reliable source of energy. However, the rapid and unchecked growth of human civilization and industrialization has led to a detrimental effect on the environment and energy resources. Fossil fuels are a finite resource, and if the present rate of consumption continues, they will likely be depleted in the coming centuries [1]. The rise in carbon dioxide levels in the lower atmosphere has prompted the search for ecologically acceptable clean and sustainable energy options. Carbon dioxide build-up has had a negative influence on the climate, creating catastrophic weather events such as excessive rainfall, floods, and drought [2]. Every country is responsible for improving the quality of its energy supplies and, when feasible, replacing non-renewable fossil fuels like coal and oil with renewable sources like wind, solar, and other kinds of energy. This approach has the potential to reduce the negative environmental consequences of carbon dioxide emissions [3]. In order to address the issue of depleting natural resources and environmental deterioration, future technologies must incorporate sustainable development concepts and criteria throughout their technological processes, products, and operations. This notion has been recognized globally, and many initiatives have been established to promote the integration of these concepts. Sustainable development is not only beneficial for the environment but also for society and the economy in the long term. By merging environmental and social considerations into the design and production of new technologies, it is possible to create products and services that are not only eco-friendly but also efficient and cost-effective.

One of the key components of sustainable development is to minimize waste and pollution throughout the life cycle of the product. This can be achieved by using renewable resources, reducing energy consumption, and maximizing the product's lifespan. Furthermore, sustainable technologies offer opportunities for creating new industries and jobs, improving public health, and creating more equitable societies. Wind energy, which has been used by humans for millennia for purposes such as sailing, grinding grains, and pumping water, is a feasible form of renewable energy that can help mitigate the negative environmental consequences of non-renewable energy. However, it was not until the late 19th century that wind power was harnessed to generate electricity [4]. Since then, wind energy technology has undergone significant advancements, making it one of the rapidly developing renewable energy sources worldwide. The earliest recorded use of wind energy was in Persia (present-day Iran) around 5,000 BC. Persians used wind energy to power their sailing ships, grind grains, and pump water. Similarly, ancient Egyptians and Greeks also used wind power to propel their ships. However, it was not until the 12th century that the first windmill was developed in Europe. These windmills were used primarily to grind grains and pump water. Windmills continued to evolve and become more efficient, and by the 19th century, they had become a common sight across Europe and North America [5].

The first electricity-generating wind turbine was developed by Charles Brush in Cleveland, Ohio, in 1887. Brush's wind turbine had a rotor diameter of 17 m and generated around 12 kW of electricity. However, it was not until the 1940s that the first large-scale wind turbines were developed in the United States. These turbines had a rotor diameter of around 30 m and generated up to 100 kW of electricity. In the 1950s, wind energy began to be used to power remote locations such as farms and ranches.

Wind energy technology has undergone significant advancements in recent decades, making it a cost-competitive and reliable source of electricity. In addition to technological advancements in wind turbines were various significant developments in wind farm design and operation. Wind farms are designed to take advantage of the prevailing wind patterns in a particular region, and turbines are placed strategically to maximize energy production while minimizing environmental impacts. Wind farm operators use advanced software to monitor and control the turbines, allowing them to optimize energy production and minimize downtime [6].

1.2 Historical Background

Wind energy has been used throughout human history from ancient times, long before coal and refined petroleum were discovered. Notably, the Egyptians used windmills for water pumping some 3,000 years ago, while Chinese farmers used vertical-axis wind turbines to empty rice fields several centuries before their European counterparts. The horizontal-axis windmill is said to be originated in Europe and was first mentioned in the Duchy of Normandy around 1180 [7]. The generation of electricity using windmills began in 1887 when Prof. James Blyth built a windmill in Scotland for this purpose. In 1888, Bruch and his colleagues successfully developed a wind machine that was put into operation on the Atlantic coast. Wind power technology continued to evolve and develop over time. Kurt Bilau, in 1920, incorporated the Ventikanten blade, which utilized an aircraft air-foil developed by him and Betz, into modern windmill designs. Small wind machines (less than 1 kW) and windmills without an electrical system proliferated in rural areas of the United States during the 1920s and 1930s. The stages of wind technology development are given in Table 1.1. With almost 600,000 units erected at this time, windmill use in the US reached its height in popularity. Developed in the US for the first time in 1941, the contemporary horizontal-axis wind turbine was primarily utilized to supply electricity to farms without access to power lines. However, from the 1950s onward, the market for wind turbines slowly started to decline due to the widespread expansion of electric power lines [8].

Table 1.1 Stages of wind technology development.

Time period	Key developments in wind energy technology
Ancient times	Windmills used for milling grain and pumping water
Late 1800s	Charles F. Brush builds the first large-scale wind turbine in Ohio, USA
1920s–1930s	First electricity-generating wind turbines developed in Europe
1940s–1950s	Utility-scale wind turbines developed for remote areas with no access to a power grid
1970s	Advancements in aerodynamics and materials lead to more efficient wind turbines
1980s	Wind power begins to gain popularity as a viable source of renewable energy
1990s	Introduction of variable-speed turbines and the use of power electronics to improve efficiency
2000s	Continued improvements in turbine design, including larger rotor diameters and higher hub heights
Present day	Advancement of offshore wind farms and increasing use of wind energy storage technologies

1.3 Use of Wind Energy in Specific Countries

Wind energy has rapidly come into utilization in the US since the 1980s, with tax breaks playing an important role in stimulating investment. As a result, the price of wind-generated power has dropped significantly from 35 to 4 cents/kWh. The Stateline Wind farm, which is now under development, will be the biggest in the world. Furthermore, wind power capacity in the US has increased rapidly [9].

However, with an average annual growth rate of 22% over the previous 6 years, Europe has overtaken all other regions as the world leader in wind energy. The European Wind Energy Association has increased its estimates for Europe's wind capacity in 2010, and offshore projects are starting to materialize off the shores of numerous European nations [10].

Germany, which in 1997 outperformed the United States in terms of wind capacity, has since emerged as Europe's future trendsetter. Denmark, which has the third-largest wind energy and is the top manufacturer and exporter of wind turbines, intends to source 50% of its domestic energy from wind by 2030. Other nations with large increases in wind capacity include Spain, Canada, India, and Japan, all of which have set goals or incentives to promote future expansion [11].

1.4 Wind Technology

Wind power can be harnessed directly as mechanical power or indirectly as electrical energy through the conversion of wind's kinetic energy (KE). All wind energy systems must include a wind turbine because it transforms the KE of the wind into mechanical power which can be used for various tasks. The wind turbine generates electricity by using the wind's energy to move a rotor, which powers a generator.

In the early 1900s, the first wind turbine intended to produce energy was developed. Since then, substantial technological and design developments have improved the efficiency and dependability of wind turbines. Today, various varieties and sizes of wind turbines exist, from small ones used to power homes and small businesses to large ones used to produce electricity on a large scale for utilities. While wind turbine technology has been gradually improving, there have been remarkable advancements in turbine design [12]. In particular, modern innovations and optimizations in turbine technology have led to significant improvements in both power output and efficiency. Additionally, the development of specialized generators and the use of power electronic devices have enabled gearless turbine designs. Currently, wind turbines are typically divided into two main types: horizontal-axis turbines (HAWTs) [13] and vertical-axis turbines (VAWTs). HAWTs are more commonly used in the wind industry due to their higher efficiency compared to VAWTs [14].

1.4.1 Wind Energy Conversion System (WECS)

The wind turbine serves as the primary part of the WECS and facilitates the conversion of KE into mechanical energy. A coupling device gear train connects the wind turbine to an electrical generator [16]. To prevent disruptions and safeguard the system or network, an appropriate controller is used to transmit the generator's output to the electrical grid [17]. Wind turbines, which employ specifically created blades to convert the KE of the wind into rotational energy, are employed to capture wind energy. The generator uses this mechanical energy to turn it into electrical energy. The generator is linked to the rotating blades *via* a shaft, allowing mechanical energy to be transferred. The layout of a wind energy conversion system is shown in Figure 1.1.

The generator's output may either be used to power a stand-alone load, such as a house or a small town, or it can be fed into the grid through a transformer. Specialized knowledge and skill in the domains of aerodynamics, mechanics, electronics, and control systems are necessary to enable the efficient and effective conversion of wind energy [18, 19]. Such abilities may be used to optimize wind energy systems so that they produce the most electricity possible while having the least negative environmental effect.

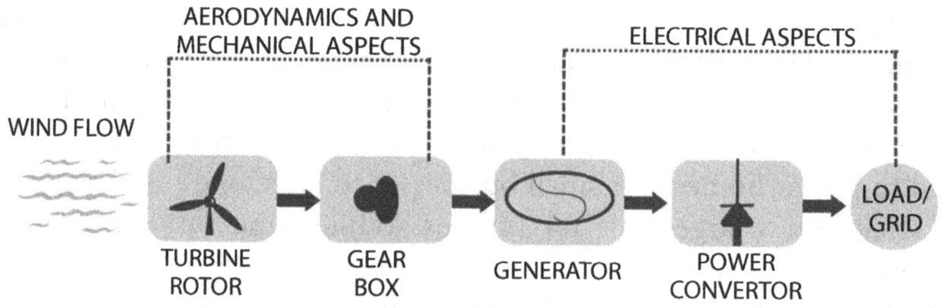

Figure 1.1 Layout of a wind energy conversion system [15].

1.4.2 Electric Generator

Due to their high dependability and effective energy collection, permanent magnet synchronous generators (PMSG) and double-fed induction generators (DFIG) are becoming more and more popular as the preferred generators in WECS. Synchronous generator-powered wind turbines are frequently referred to as gearless or direct-drive wind turbine generators [20, 21]. The PMSG has lately attracted interest since it creates a magnetic field using permanent magnets and is self-exciting without the need for a DC source. Induction generators, meanwhile, are frequently employed because of their brushless design, excellent dynamic response, straightforward operation, and reduced cost [22]. The size, power output, and operational needs of the wind turbines are only a few of the variables that affect the generator technology choice. To increase their effectiveness, dependability, and overall performance, WECS generator technologies have undergone extensive research and growth in recent years. The broad adoption of wind energy will be greatly aided by the progress of new generator technologies as the need for renewable energy keeps rising. However, they require reactive power to generate a magnetic field, which is drawn from the grid in grid-connected applications, but in standalone applications, a power electronics converter or capacitor bank is employed to provide reactive power. A self-excited induction generator (SEIG) is created when a capacitor is connected to an induction generator [23]. Figures 1.2 and 1.3 represent the general outline of a power converter.

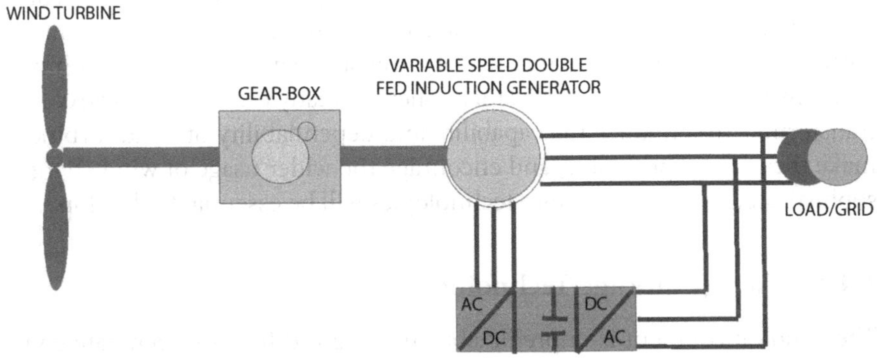

Figure 1.2 Variable-speed synchronous/induction generator with full-scale power converter [24].

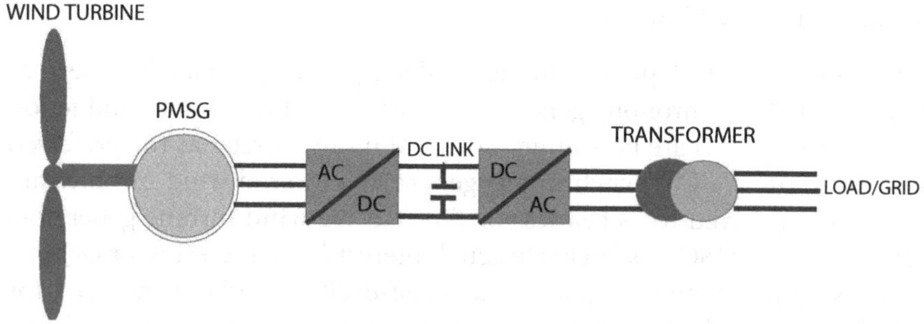

Figure 1.3 Variable-speed, gearless multi-pole PMSG-based WECS [24].

1.4.3 Evolution of Power Electronics

The aim of power electronics in WECS systems is to improve reliability, energy efficiency, and performance while reducing mechanical stress. As a result, the complete system may connect more fluidly with the grid and function as an adjustable power producing unit. Power electronics are especially important in variable speed wind turbine systems, as they allow for better control over the generator. In fixed speed wind turbine systems, thyristors can act as soft starters. Although expensive, power electronics can significantly simplify the mechanical design of wind turbines, reduce mechanical stress, and lessen the effects of wind gusts. Additionally, they can do away with the need for gearboxes, which are a frequent cause of failure and losses in wind turbines. Researchers are creating advanced power electronic technologies that can convert power at greater voltage levels to fulfil the rising demand for wind energy. As an interface for a full-scale power converter, galvanically separated DC–DC converters using a cascaded H-bridge converter architecture with a back-to-back arrangement are one possible alternative [25]. This concept makes use of a transformer that operates within a medium-frequency range of several kilohertz, greatly reducing the size of the transformer. To enhance the capability and dependability of wind turbines, make them more profitable, and encourage the wider usage of wind energy, sophisticated power electronics technologies will be essential to develop.

1.4.4 Energy Storage Technology

The wind's movement is unpredictable, making it difficult to generate consistent power. Unstable wind power can create significant issues for power system operation, stability, and planning. Wind energy requires the efficient control and use of intermittent wind power. During times of little or no wind availability, excess wind-generated electricity may be stored and retrieved by means of energy storage technologies. Numerous storage technologies have been thoroughly investigated, including batteries, compressed air energy storage (CAES), flywheels, and pumped hydro storage (PHS). These technologies have distinctive qualities, benefits, and drawbacks. Careful attention must be given to elements including energy density, efficiency, scalability, environmental impact, and cost-effectiveness when choosing an appropriate storage system for wind energy. To ensure dependable wind energy production, an energy storage system (ESS) is needed [26]. In WECS, extra electrical energy is stored in the ESS and then delivered to the load in the event of a power outage. Batteries are the most often utilized energy storage technology in this context, despite the existence of other energy storage technologies, due to their capacity to store energy as electrochemical energy [27].

1.5 Horizontal-Axis Wind Turbines (HAWTs)

Since the first windmill was built in ancient Persia, the wind turbine industry has advanced significantly. The most common and largely utilized form of wind turbine in modern times is the HAWTs. These turbines consist of a rotor with many horizontally revolving blades that face the wind. The wind's kinetic energy causes the blades to begin rotating, which, in turn, powers a generator to generate electricity. HAWTs are favored in the wind energy sector because of their exceptional efficiency, aerodynamic design, and capacity to capture wind from many directions [28]. In this chapter, we will discuss the history, design, components, working principles, and applications of HAWTs.

1.5.1 History

The first horizontal-axis wind turbine was made in Scotland in 1887 by Professor James Blyth. It had a 10-m-diameter rotor and generated electricity to light his cottage. Later, in the 1930s, a Danish engineer, Poul La Cour, developed HAWTs that were used to power homes and farms. In the 1970s, the oil crisis led to a surge of interest in wind power, which accelerated the development of modern HAWTs [29, 30].

1.5.2 Design

A tower, nacelle, and rotor are the components of a HAWT system. The hub of the rotor, which is attached to a main shaft, is made up of two or three blades. The gearbox, generator, and other components that change the rotor's rotational energy into electrical energy are housed in the nacelle, which also acts as a containment. The tower supports the rotor and nacelle and can be made of steel, concrete, or hybrid materials [31, 32].

1.5.3 Components

The blades of HAWTs are typically made of fiberglass, carbon fiber, or other composite materials. The hub and main shaft are made of steel, while the gearbox is usually made of cast iron or aluminum [33]. The generator can be a conventional synchronous generator or a newer technology such as a permanent magnet generator. The control system of the turbine includes sensors, actuators, and a controller that regulates the pitch of the blades and the speed of the rotor [34].

1.5.4 Working Principle

HAWTs work on the concept of the aerodynamic lift principle. The turbine's blades are made to generate a pressure differential between their top and lower surfaces when the wind blows over them, which causes the blades to rotate.

The primary shaft, which rotates in reaction to the motion of the blades, is connected to the generator. The mechanical energy generated by the blades' spinning is subsequently changed into electrical energy with the use of a generator. The power grid subsequently receives this electrical energy for distribution and utilization. HAWTs use complex fluid dynamics and mechanics to convert wind energy into electrical energy, which calls for careful design and engineering to maximize the efficiency of the turbine. HAWTs have emerged as a significant renewable energy source in recent years, and addressing the rising need for sustainable energy sources will depend on their continuing growth and improvement.

1.5.5 Applications

HAWTs are used to generate electricity for a variety of applications, including residential, commercial, and industrial. They can be installed onshore or offshore, depending on the availability and quality of wind resources. HAWTs can also be combined with other renewable energy sources such as solar power and energy storage systems to provide a stable and reliable power supply.

1.6 Vertical Axis Wind Turbine (VAWT)

VAWT are often used in homes as a sustainable energy source. It is made up of a rotor shaft and two or three vertically moving blades that mimic the action of spinning coins on their edges. Unlike standard turbines, the generator is located at the bottom of the tower, with the blades around the shaft [35].

1.6.1 Working Principle

The vertically oriented blades of the vertical-axis wind turbine, which revolve around a vertical shaft and use wind energy to generate electricity, are essential to the device's operation. The generator through which the rotor is attached to changes the mechanical energy generated by the wind into electrical energy. The blade, shaft, bearing, frame, and blade support are among the turbine's crucial parts [36].

Displayed below is a block diagram depicting the components of a VAWT. The energy produced by this turbine can be utilized by any form of load, with an automatic lighting system serving as an example. This diagram (Figure 1.4) consists of a VAWT, gearbox, generator, battery, LDR circuit, and LED.

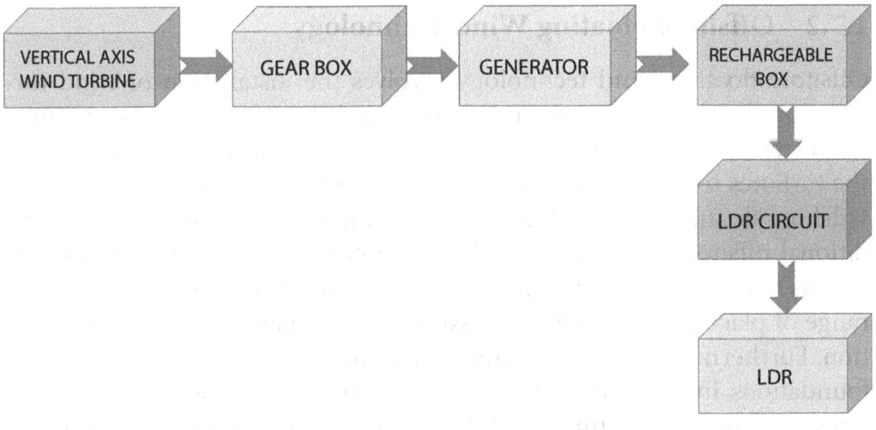

Figure 1.4 Block diagram depicting the components of a vertical-axis wind turbine.

1.7 Current Technologies in Wind Power Generation

Airborne wind energy is an auspicious technology that employs tethered airborne devices to exploit wind power. These airborne machines, also known as kite turbines or airborne wind turbines, generate electricity by flying in certain patterns that allow them to collect the energy of the wind at greater heights [37]. This electricity may be used for a variety of reasons, including powering homes, businesses, and remote places without access to traditional power systems. Airborne wind energy has various advantages over regular wind turbines, including a lower environmental effect, lower installation and maintenance costs, and the ability to generate power in regions where standard wind turbines are infeasible [38, 39].

1.7.1 Buoyant Airborne Turbine (BAT)

BAT is a technology that uses wind energy *via* a wind turbine mounted to a helium-filled blimp-like object. This contraption can float up to 2,000 ft in the air, capturing strong and persistent winds. The wind moves the turbine's blades, creating power that is delivered to the ground through a tether. BAT offers various benefits over standard wind turbines, including the capacity to operate at higher elevations, where winds are more consistent and forceful, and the potential to be employed in isolated and offshore sites. Furthermore, BAT takes less material to construct and less area to operate, resulting in a lower environmental effect.

1.7.2 Offshore Floating Wind Technology

Offshore floating wind technology involves the installation of wind turbines on floating platforms in deep oceans distant from the shore. These platforms, sometimes known as floating wind farms, use cables to tether the turbines to the seabed and transfer the power generated back to land. Offshore floating wind technology has a number of advantages over traditional offshore wind turbines. It can capture stronger winds in deeper oceans where standard turbines cannot, and it can be utilized in a larger range of places, expanding the possibilities for renewable energy production. Furthermore, because floating wind farms do not require permanent foundations in the seabed, they have a lower environmental effect than typical offshore wind turbines [40]. Offshore floating wind, while being a relatively new technology, has displayed outstanding output in terms of efficiency, dependability, and cost-effectiveness and is predicted to perform a significant part in the future of renewable energy [38].

1.8 Advantages

VAWTs have a number of benefits, including worker safety and the ability to produce power from wind that is blowing in any direction. Their gearbox, generator, and other components are positioned on the ground, making them expandable and not needing a sturdy supporting tower. They are simpler to install and more cost-effective to design than horizontal-axis turbines. Additionally, they are transportable, making it simple to relocate them from one place to another. These turbines pose less of a threat to people and birds because they have less speed blades. They can function in any weather, even in steep areas with changing winds. They can also be placed close to the ground, which lowers the cost of construction and maintenance. They can operate quietly in residential areas because of their straightforward design. Additionally affordable, efficient, and silent, VAWTs are a perfect renewable energy source for residential use, especially in metropolitan settings [41]. A list of countries utilizing wind power as a source of electricity is given in Table 1.2.

Table 1.2 List of countries utilizing wind power as a source of electricity.

Country	Wind power capacity	Percentage of electricity from wind power
China	281.5 GW	10.4%
United States	127.2 GW	8.4%
Germany	65.8 GW	27.2%
India	40.5 GW	5.6%
Spain	28.4 GW	22.7%
United Kingdom	24.2 GW	24.2%
France	17.8 GW	8.1%
Brazil	18.8 GW	9.3%
Canada	16.5 GW	8.5%
Italy	12.8 GW	9.9%

1.9 Disadvantages of Wind Energy

Before relying solely on wind energy, it is vital to consider its drawbacks. For starters, storing energy during peak output periods could be costly. Second, wind energy is unreliable due to its unpredictability. Furthermore, wind farms necessitate large open places for construction since wind turbines might have a negative visual effect on the surrounding area. Wind turbines also contribute to noise pollution, which is an issue. Furthermore, wind energy can be generated only in areas with strong, consistent winds, which may be remote from areas with high electricity demand, raising the cost of transmission. Furthermore, because wind turbines have a lower average efficiency than fossil fuel generation, a greater number of turbines may be necessary to achieve equal results.

1.10 Conclusion

Wind power has evolved significantly technologically over time, from its early use for grinding and pumping to the current large-scale energy generation *via* wind turbines. These developments in materials science, aerodynamics, and computer control systems have considerably increased wind energy harvesting efficiency and cost-effectiveness. Wind energy is now a crucial component of the global energy mix, and it is well positioned to play a critical part in the transition to a low-carbon, sustainable future. Given the current rate of innovation and investment in the field, wind energy is very likely to advance and emerge as a critical source of renewable energy for future generations.

References

1. Ma, Q., Lu, H., Wind energy technologies integrated with desalination systems: Review and state-of-the-art. *Desalination*, 277, 274, 2011.
2. Aman, M.M., Jasmon, G.B., Ghufran, A., Bakar, A.H.A., Mokhlis, H., Investigating possible wind energy potential to meet the power shortage in Karachi. *Renew. Sustain. Energy Rev.*, 18, 528, 2013.
3. Madlool, N.A., Saidur, R., Rahim, N.A., Islam, M.R., Hossian, M.S., An exergy analysis for cement industries: An overview. *Renew. Sustain. Energy Rev.*, 16, 921, 2012.
4. Komarov, D., Stupar, S., Simonović, A., Stanojević, M., Prospects of wind energy sector development in Serbia with relevant regulatory framework overview. *Renew. Sustain. Energy Rev.*, 16, 2618, 2012.

5. Jacobsson, S., Johnson, A., The diffusion of renewable energy technology: An analytical framework and key issues for research. *Energy Policy*, 28, 625, 2000.
6. Amano, R.S., Review of Wind Turbine Research in 21st Century. *J. Energy Resour. Technol. Trans. ASME.*, 139, 0508011, 2017.
7. Saffir, H.S., Asce, F., Florida's approach to hurricane-resistant design and construction. *J. Wind Eng. Ind. Aerodyn.*, 32, 221, 1989.
8. Sørensen, B., A history of renewable energy technology. *Energy Policy*, 19, 8, 1991.
9. Saidur, R., Islam, M.R., Rahim, N.A., Solangi, K.H., A review on global wind energy policy. *Renew. Sustain. Energy Rev.*, 14, 1744, 2010.
10. Wang, Q., Zhan, L., Assessing the sustainability of renewable energy: An empirical analysis of selected 18 European countries. *Sci. Total Environ.*, 692, 529, 2019.
11. Papież, M., Śmiech, S., Frodyma, K., Factors affecting the efficiency of wind power in the European Union countries. *Energy Policy*, 132, 965, 2019.
12. Joselin Herbert, G.M., Iniyan, S., Sreevalsan, E., Rajapandian, S., A review of wind energy technologies. *Renew. Sustain. Energy Rev.*, 11, 1117, 2007.
13. Demirbas, A., Electrical power production facilities from green energy sources. *Energy Sources, Part B Econ. Plan. Policy*, 1, 291, 2006.
14. Drewry, M.A., Georgiou, G.A., A review of NDT techniques for wind turbines. *Insight Non-Destructive Test. Cond. Monit.*, 49, 137, 2007.
15. Devashish, Thakur, A., Panigrahi, S., Behera, R.R., A review on wind energy conversion system and enabling technology. In: *International Conference on Electrical Power and Energy Systems*, ICEPES 2016. pp. 527–532, 2017.
16. Cheng, M., Zhu, Y., The state of the art of wind energy conversion systems and technologies: A review. *Energy Convers. Manag.*, 88, 332, 2014.
17. Blaabjerg, F., Teodorescu, R., Liserre, M., Timbus, A. V., Overview of control and grid synchronization for distributed power generation systems. *IEEE Trans. Ind. Electron.*, 53, 1398, 2006.
18. Polinder, H., Van Der Pijl, F.F.A., De Vilder, G.J., Tavner, P.J., Comparison of direct-drive and geared generator concepts for wind turbines. *IEEE Trans. Energy Convers.*, 21, 725, 2006.
19. Cotrell, J.R., A preliminary evaluation of a multiple-generator drivetrain configuration for wind turbines. In: *ASME 2002 Wind Energy Symposium, WIND2002.* pp. 345–352. ASMEDC, 2002.
20. Bozhko, S. V., Blasco-Giménez, R. V., Li, R., Clare, J.C., Asher, G.M., Control of offshore DFIG-based wind farm grid with line-commutated HVDC connection. *IEEE Trans. Energy Convers.*, 22, 71, 2007.
21. Chalmers, B.J., Wu, W., Spooner, E., An axial-flux permanent-magnet generator for a gearless wind energy system. *IEEE Trans. Energy Convers.*, 14, 251, 1999.
22. Abdel-Khalik, A.S., Ahmed, S., Massoud, A.M., Elserougi, A.A., An improved performance direct-drive permanent magnet wind generator using a novel single-layer winding layout. *IEEE Trans. Magn.*, 49, 5124, 2013.
23. Yan, H., Liu, Z., Qi, R., A review of humidity gradient-based power generator: Devices, materials and mechanisms. *Nano Energy*, 101, 107591, 2022.

24. Hossain, M.M., Ali, M.H., Future research directions for the wind turbine generator system. *Renew. Sustain. Energy Rev.*, 49, 481, 2015.
25. Yamazaki, K., Nishioka, K., Shima, K., Fukami, T., Shirai, K., Estimation of assist effects by additional permanent magnets in salient-pole synchronous generators. *IEEE Trans. Ind. Electron.*, 59, 2515, 2012.
26. Tavner, P.J., Xiang, J., Spinato, F., Reliability analysis for wind turbines. *Wind Energy*, 10, 1, 2007.
27. Geng, H., Yang, G., Linear and nonlinear schemes applied to pitch control of wind turbines. *Sci. World J.*, 2014, 1, 2014.
28. Johnson, D.A., Gu, M., Gaunt, B., Wind Turbine Performance in Controlled Conditions: BEM Modeling and Comparison with Experimental Results. *Int. J. Rotating Mach.*, 2016, 1, 2016.
29. Manwell, J.F., McGowan, J.G., Rogers, A.L., Wind Energy Explained: Theory, Design and Application. Wiley 2010.
30. Beurskens, J., The History of Wind Energy. In: *Understanding Wind Power Technology: Theory, Deployment and Optimisation*, pp. 1–44, 2014.
31. Rehman, S., Mahbub Alam, M., Alhems, L.M., Mujahid Rafique, M., Horizontal AxisWind Turbine Blade Design Methodologies for Efficiency Enhancement A Review. *Energies*, 11, 506, 2018.
32. Benini, E., Toffolo, A., Optimal design of horizontal-axis wind turbines using blade-element theory and evolutionary computation. *J. Sol. Energy Eng. Trans. ASME.*, 124, 357, 2002.
33. Suresh, A., Rajakumar, S., Design of small horizontal axis wind turbine for low wind speed rural applications. *Mater. Today Proc.*, 23, 16, 2019.
34. Lee, D., Hodges, D.H., Patil, M.J., Multi-flexible-body Dynamic Analysis of Horizontal Axis Wind Turbines. *Wind Energy*, 5, 281, 2002.
35. Kumar, R., Raahemifar, K., Fung, A.S., A critical review of vertical axis wind turbines for urban applications. *Renew. Sustain. Energy Rev.*, 89, 281, 2018.
36. Pope, K., Dincer, I., Naterer, G.F., Energy and exergy efficiency comparison of horizontal and vertical axis wind turbines. *Renew. Energy*, 35, 2102, 2010.
37. Ali, Q.S., Kim, M.H., Unsteady aerodynamic performance analysis of an airborne wind turbine under load varying conditions at high altitude. *Energy Convers. Manag.*, 210, 112696, 2020.
38. Liu, Y., Li, S., Yi, Q., Chen, D., Developments in semi-submersible floating foundations supporting wind turbines: A comprehensive review. *Renew. Sustain. Energy Rev.*, 60, 433, 2016.
39. Trevisi, F., McWilliam, M., Gaunaa, M., Configuration optimization and global sensitivity analysis of Ground-Gen and Fly-Gen Airborne Wind Energy Systems. *Renew. Energy*, 178, 385, 2021.
40. Li, L., Liu, Y., Yuan, Z., Gao, Y., Dynamic and structural performances of offshore floating wind turbines in turbulent wind flow. *Ocean Eng.*, 179, 92, 2019.
41. Roga, S., Bardhan, S., Kumar, Y., Dubey, S.K., Recent technology and challenges of wind energy generation: A review. *Sustain. Energy Technol. Assess.*, 52, 102239, 2022.

2

Environmental Consequences of Wind Energy Technologies

Adarsh Kumar Arya and Ashish Kapoor*

Department of Chemical Engineering, Harcourt Butler Technical University, Kanpur, Uttar Pradesh, India

Abstract

In light of rising public interest in environmental and social problems, governments throughout the globe are evaluating the impacts of pollution on the energy paradigm. Power plants employ sustainable energy sources, including wind, solar, and hydro, to lessen their environmental impact. Wind energy, one of these global commons, has recently emerged as a promising new energy source. Wind energy is becoming a more significant component of the power sector when there is increasing anxiety over the expenses and long-term ecological effects of using petroleum-based fuels and nuclear energy. Because of its low environmental impact, wind power has received government backing through subsidies and tax credits. However, the development of wind projects has been hampered by fears of some adverse effects from onshore and offshore wind farms. The concerns are due to the possibility of major effects on birds and species of bats and marine animals as well as direct and indirect impact on humans that may halt, scale down, and significantly delay the progress of projects. The chapter investigates and discusses several elements of the environmental impacts caused by wind energy technology.

Keywords: Renewable energy, wind power, pollution, fossil fuel, environmental impact, green technology

2.1 Introduction

Energy is a crucial component of our daily life and is a fundamental requirement for the advancement of civilization. Modern society and the global

**Corresponding author*: ashishkapoorchem@gmail.com

Inamuddin, Tariq Altalhi and Mohammad Luqman (eds.) Wind Energy Storage and Conversion: From Basics to Utilities, (17–32) © 2024 Scrivener Publishing LLC

economy depend on reliable and renewable energy sources. However, modern civilization's rapid and unpredictable advancements and industrialization severely impact the environment and energy resources [1–5]. Reduced reliance on nuclear and fossil fuel power, reduced environmental risks, and the possibility of cheaper energy generation all contribute to the attractiveness of renewables [6]. Sustainable development concepts and criteria execution in technological processes, goods, and operations are essential if we are to halt the consumption of resources and the deterioration of our environment in the future [7–12].

Governments throughout the globe have declared targets to increase the usage of renewable energy sources. Among other renewable sources, wind energy has emerged as one of the most rapidly expanding renewable energy sources, making important contributions to the world's energy supply. Renewable energy sources like wind power are the key to a low-carbon future. The projections estimate that wind energy could contribute up to nearly 20% of the entire United States electricity by 2030 [13].

According to the published reports [6], there was a record growth of 273 TWh (17%) in wind-generated power in 2021. This increase was the largest of all renewable energy sources, 55% larger than the rise expected in 2020. A record-breaking increase in wind capacity growth enabled this rapid rise; in 2020, wind power capacity reached 113 GW, up from 59 GW in 2019. The Net Zero Emissions by 2050 scenario predicts that wind power will account for over 7,900 TWh of electricity output in 2030. Figure 2.1 shows the cumulative installed capacity of wind power, and Table 2.1 shows the onshore wind installed capacity of important European countries as of 2019.

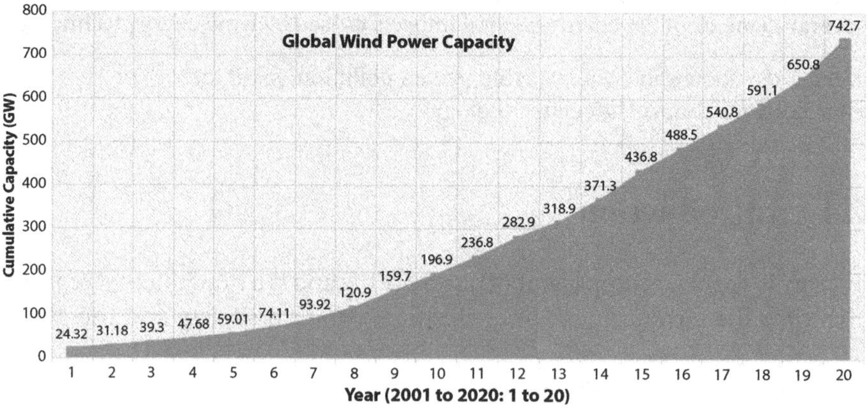

Figure 2.1 Global wind power cumulative capacity (GW) (2006–2019). Reproduced from [15] under Creative Commons CC-BY license.

Table 2.1 Onshore installed wind power of representative countries in Europe. Data from [16].

S. no.	Country	Power (MW)	S. no.	Country	Power (MW)
1	France	523	8	Netherlands	83
2	Denmark	6	9	Portugal	57
3	Sweden	459	10	Ireland	51
4	Croatia	10	11	Bosnia	36
5	Poland	17	12	Turkey	229
6	Russia	50	13	Spain	148
7	Belgium	72	14	Greece	201

The projection assumes that yearly capacity additions would increase to about 250 GW. This continuous capacity increase requires far more work, with permission for onshore wind being the most important area for improvement, followed by lowering expenses for offshore wind [14].

However, government agencies have had concerns about the successful use of wind energy. The purpose of this study is to review and summarize the issues that arise due to the incorporation of wind power generation into the networks. Figure 2.2 shows the environmental impact of using wind energy technologies on the environment.

Wind energy technologies have both positive and negative impacts on the environment. On the positive side, wind energy is a renewable source that reduces greenhouse gas emissions, mitigates climate change, and helps

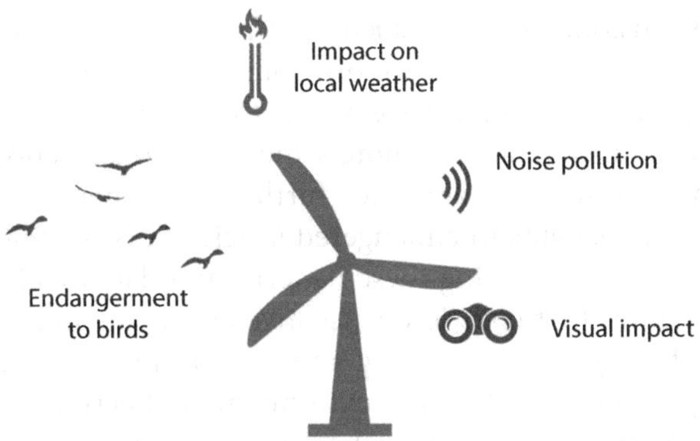

Figure 2.2 Impact of wind energy technologies on the environment.

diversify the energy mix. However, wind farms, while beneficial, require land and can potentially disrupt natural habitats, leading to endangerment of wildlife populations. Birds may collide with the rotating blades of wind turbines, necessitating careful site selection and the use of avian deterrents to minimize such risks. Additionally, wind turbines can have an effect on local weather patterns by altering wind flow in their vicinity. Noise pollution is another concern, as the spinning blades generate noise that can impact nearby residents [17]. Moreover, wind turbines can have visual impacts on the landscape, which some find unsightly or disruptive to the natural aesthetics of an area. Wind energy's intermittency necessitates complementary energy sources or storage systems. Additionally, manufacturing and installation have environmental consequences, but the energy payback period is relatively short. Proper maintenance practices are crucial. Overall, wind energy is considered more environmentally friendly than fossil fuels, but careful planning and monitoring are necessary to maximize its benefits and minimize any adverse effects. A holistic analysis is essential to give the technologists a comprehensive understanding of the subject. It also equips policy makers with all relevant information, allowing them to make more informed decisions and implement more effective, long-term solutions to the problems associated with wind power's incorporation into electrical grids.

2.2 Impact of Wind Energy on the Environment

The effects of wind-generating infrastructures (such as turbines, towers, and transmission lines) on wildlife have been studied and recorded for decades, but the data are not always readily available to the public. Despite the substantial uncertainties connected with environmental issues restricting expansion, the established land-based wind industry and the more immature offshore wind sectors continue to expand [18, 19]. In this scenario, scientists would have access to data on the environmental impacts of both onshore and offshore wind farm construction. They will be in a stronger position to argue for further investment in wind power and highlight the benefits to endangered species. It is possible to use the insights gained from analyzing these patterns and data to adjust machinery and procedures that have fewer negative effects on animals and their habitats, which may aid in the siting and consenting processes. Agencies like "Working Together to Resolve Environmental Effects of Wind Energy" (WREN) are online communities where people work together to spread information and promote the global understanding of wind energy.

Researchers, authorities, designers, and key stakeholders connect and interact *via* the WREN Hub to comprehend the possible implications of wind energy on animals, ecosystems, and ecological processes [20]. White papers and other anticipated products are only two examples of how the WREN Hub facilitates collaboration to support WREN's global operations.

The WREN Hub is located on Tethys. Tethys was created and is maintained by the Pacific Northwest National Laboratory in the United States to gather and distribute scientific data on the outcomes of wind and marine renewable energy projects. The collection on Tethys is always being expanded with the help of journal papers, technical papers, demonstrations, and other media items contributed by the WREN Hub. Such comprehensive knowledge base has over 3,500 carefully selected papers, of which over 2,300 are especially pertinent to wind power.

New Tethys content, advancements, and upcoming events are highlighted in user updates published every 2 weeks. A quarterly recorded webinar series, as are recordings of previous webinars, is also available. According to the metrics for how often researchers utilize Tethys resources, the WREN Hub encourages sharing new findings and reduces duplicate efforts to study important themes [20]. The WREN Hub remains a dependable resource that aims to increase the knowledge on how wind energy growth affects wildlife. It alleviates worries and settles disagreements over expanding this environmentally friendly energy source. To sustainably expedite the wind power sector's development as countries switch to renewable energy sources, widespread information sharing and collaborative activities are essential.

2.3 Key Environmental White Paper Issues Related to Wind Power

White papers are one of WREN's core operations, and they are written on issues of significant importance to WREN participants but about which there is little information in the current public literature. These white papers address a wide range of subjects related to environmental management, including adaptive management (AM), environmental risk management, and environmental trade-offs. WREN participants have agreed that these are significant areas in which the current academic literature is deficient. Every year until 2020, a minimum of one white paper was released. When all papers are finished, a report is written on how they relate. These articles appeared in WREN and other scholarly publications.

Many different kinds of development have employed AM, a learning-based management strategy [21], to decrease scientific uncertainty. AM has been recognized as a means to improve risk management in the wind energy sector by allowing monitoring and mitigation strategies to evolve with each new project. This is accomplished through hypothesis-driven data collection. However, AM use in wind energy production has been restricted; the United States is the primary location, while other nations have adopted some of the same practices [22]. The WREN nations have published a white paper to understand better how AM concepts are used in the wind power sector in different countries. Implementing AM in wind energy development might not be easy because it requires a versatile procedure that does not impede project funding or slows the permitting procedure.

For instance, when power purchase contracts are signed, it may not be easy to establish an adaptive mechanism that allows or mandates developers to scale down, alter operations, or implement mitigation measures. Constant monitoring expenses and potential revenue loss from necessary mitigation measures like operating restrictions are significant issues for wind power stations implementing AM [23]. The absence of laws or regulations that mandate and define AM and the dearth of supporting resources only adds to the difficulty of implementing it consistently. Most WREN nations have neither an official usage for AM nor any legislation or regulations. However, in certain countries, AM or the implementation of its principles is explicitly included in natural resource law, regulation, and recommendations for the construction of wind energy projects. It was discovered that AM had been used in several wind energy projects. However, numerous important animal species may be impacted by wind energy at far larger geographical and chronological ranges than those of a single project site. Because of this scale mismatch, it is difficult for a single initiative to contribute effectively to lowering scientific uncertainty and fostering iterative learning. For AM to be most successful, it has to be implemented on a broader scale than individual projects in terms of space and time.

Scientific uncertainty may be reduced by fine-tuning AM's utilization of geographical and temporal scales. Using environmental impact data from ongoing projects to inform the preparation and oversight of future projects requires establishing institutional procedures and institutions within national or provincial regulatory authorities. Scientific uncertainty may be decreased, and lessons learnt will guide the licensing and management of new wind energy production globally if AM for wind energy projects is to be enhanced.

2.4 Individual Effects on Population Impacts

There is much evidence documenting the negative effects of wind-generating installations on wildlife all around the globe, particularly birds, bats, and marine animals. Collisions may directly cause death, but they can also have indirect impacts, such as stress on the body, the loss of a habitat, or even relocation. The cumulative impact of these factors on a population's long-term viability has to be considered. The licensing framework often looks at the consequences on people within a limited period and geographic area. However, from a conservationist's perspective, it is important to extrapolate these results to the population as a whole. The costs to wildlife must be weighed against the broader social advantages of wind energy.

Therefore, a shift in strategy and decision-making is essential. The Wind Research and Education Network (WREN) produced a white paper titled "Individual Effects to Population Impacts," highlighting and examining the methodological challenges of evaluating the impacts of wind energy installations on whole populations. Delimitation based on biogeography or genetics has scholarly merit but may be challenging to put into practice. Nonetheless, we suggest that the definition includes the species' present conservation status and past population dynamics. Another option is to use a tiered strategy to help bring together the different levels of importance [24]. Problems also arise in defining population impact, making accurate predictions, and seeing them when they occur.

There are a variety of demographic parameters that can be used as impact evaluation metrics, such as the number of people or density [25], population growth rate [26], and mortality [27]. They had a reference point to measure the change in these variables. Both temporal (as in a before–after design) and spatial (as in a control–impact design) approaches can be taken [28]. When research strategies are combined [29], the results are more reliable since they consider possible changes in the total environment. These metrics are particularly useful in the framework of risk-based decision-making, whereby the size of a change, as well as its variability and probability, may be stated [30].

Extrapolating existing data to cover bigger regions and longer periods is the most fundamental of the various statistical and modeling tools that may be used to estimate consequences during preconstruction assessment [31]. Population viability analysis uses sophisticated techniques such as matrix models of populations [32] to estimate a population's vulnerability to extinction [33]. Individual-based models have been created more recently, replicating all people's birth, death, and migration inside the model domain in discrete time steps. Despite their impressive strength, such models are computationally expensive and data-heavy. How to establish impact thresholds is the ultimate issue that has to be solved by decision-makers. Ecological, utility, and choice thresholds are the three categories that may be drawn from this concept. Population models are used to determine ecological thresholds, which are then activated by changes in system dynamics (such as when a population's growth rate becomes negative). When the significance of a management result changes, utilities are activated. The maximum sustainable harvest rate is an excellent illustration of this principle. Appropriate biological change [34] and prospective biological removal [24, 29] are methods for determining utility thresholds. Finally, decision thresholds represent the situations that should provoke a reaction from management. Ecological and utilitarian criteria will usually guide such choices.

From a policy viewpoint, it is important to consider reconciling precautionary and risk-based approaches regardless of the methodological technique employed to analyze population-level consequences and how thresholds are determined. When evaluating possible effects on species of concern, it is vital to account for the inherent ecological and demographic stochasticity in population dynamics as well as the limitations of any model. Policies that promote AM, aiming to minimize scientific uncertainty while protecting biodiversity, would encourage the adoption of population-level analyses to achieve a fair balance between the increase of wind energy and the protection of animal populations.

2.5 Comprehending the Overall Effects of Wind Power on Wildlife

Wind farms attract investment from countries operating on land and at sea. Wind energy is expanding so quickly and on such a massive scale that it is difficult to predict (much alone verify) its effects on biodiversity. Collisions are an example of direct impacts, whereas the destruction of habitat and barrier effects are examples of indirect impacts [35]. Most species-specific impact evaluations done for wind farm licensing have shown that the risks to wildlife are manageable, with or without mitigation. It is inevitable that as wind farms proliferate, the potential for animal harm will grow—for instance, although a single wind farm would not cause much trouble, an expanding number of wind farms and the infrastructure they need might significantly affect local populations or whole ecosystems. The absence of good evaluation techniques, however, means that these widespread effects on the environment are often ignored.

Social and legal acceptance of wind farm licenses requires data-driven judgements regarding ecological and economic spatial planning. However, the complexity of the cumulative impact assessment makes it challenging to collect accurate data and conduct rigorous evaluations—for instance, the temporal and geographical dimensions that should be used in cumulative effect analysis (CEA) are not certain. The question of dealing with the multiplicity of uncertainties present in a CEA has been the subject of much discussion. Furthermore, the oceanic environment makes it more difficult to gather data and understand biological processes related to offshore wind. Worldwide investigations, surveys, master plans, and administrative procedures increasingly require standardized approaches, standards, and tools to assess the impact of wind energy on the environment.

2.6 Considerations for the Environment when Making Choices

Increased wind energy production has raised environmental issues, including habitat losses, disturbances, and collisions among animal populations. Environmental protection and renewable energy are public goods, but their costs and benefits may be evaluated differently depending on the geographic scale. Concerns about the local environment and the (international) advantages of growth in terms of lowering emissions of greenhouse gases might be difficult to balance. However, environmental impacts at larger dimensions may result from fragmented growth. Even in places where wind energy seems well established, such difficulties are cropping up in various worldwide societies. The two sides in this dispute are often called "Greens." Although it should go without saying that all human acts have effects, some of which may be negative, it is not always the norm to see environmental outcomes as public goods. Understanding different solutions' possible costs and advantages is crucial for optimizing local and global environmental and climatic benefits. It may be helpful to solve these challenges to transition to greener energy sources while maintaining natural ecosystems. Implementing renewable energy to safeguard against climate change has global advantages, including intrinsic value and social and economic advantages. Decision-making processes for developing wind energy must thus consider such trade-offs, finding a happy medium between environmental protection and the implementation of wind power. The expansion of wind energy is only one industry that has a ripple effect on the natural world. Wind power, on the other hand, helps cut down on harmful emissions of greenhouse gases. Prioritization and trade-offs across sectors and geographical scales (e.g., roads against wind energy, afforestation *versus* wind energy) must be examined concurrently.

Including all public goods in some way impacted by the expansion of wind energy and other humanitarian operations in the decision-making framework and accompanying tools may be beneficial. A WREN white paper will provide an in-depth discussion of these points of view. To guarantee that the right metrics are considered for the intended result, complicated processes and systems might benefit greatly from using more sophisticated trade-offs and priority management techniques.

2.7 Wind Power and Risk Management

Risk-based management identifies potential dangers and ranks them in order of severity to minimize, monitor, and control negative consequences. This is done while accounting for the system's inherent uncertainties. Risk-based management will guide permits and operational choices in the wind energy sector, reducing the likelihood of unintended consequences like animal effects while increasing output. Risk is generally agreed to be proportional to the frequency with which an event might occur and the severity of its potential outcomes. Because of this lack of data, authorities often err on caution when deciding whether or not to green-light a wind energy project. This method might exaggerate the potential effects on wildlife or fail to account for crucial unanticipated results [36]. A WREN white paper is now in the works, and it will examine and offer ways for more accurately assessing risk at both onshore and offshore wind farms. Better risk estimations will help guide the construction of wind farms in the future and make it easier to gather and analyze data. The purpose of this white paper is not to add unnecessary complexity to existing consenting/permitting procedures for wind farms but rather to provide a foundation of information and analysis that will speed up future wind development. Permitting wind farms often takes a conservative approach, especially in untested regions, and this strategy considers that the presence of even a single animal at a wind farm would affect that species.

There is also an assumption of constant occurrence of otherwise irregular installation operations or operating modes. If regulators are not given all the facts, they may make more cautious than necessary judgments since they will be basing their judgment on inaccurate assumptions about the level of danger. The wind business is subjected to unnecessary scrutiny and must frequently spend disproportionate amounts to defend itself. Allocating resources to track and reduce detrimental behaviors, rather than the interactions that place animals at increased risk, may result in population declines in populations already experiencing pressure-induced declines. Land-based wind farms pose potential threats to animals because of construction-related disturbance, habitat loss, and operational disturbances such as blade strikes. The development of offshore wind farms may have several negative effects, including noise pollution, sedimentation shifts, altered flow patterns, and even blade strikes when they are in use.

Monitoring efforts should shift their focus from assessing the effects on individual animals to gauging the cumulative effects on populations, particularly those that are threatened or endangered and in need of special protection under national or global management policies or are under increasing stress due to other natural or anthropogenic activities.

Wildlife population risk assessment is an imprecise science due to a lack of information on life histories and the spatial and temporal distributions of different parts of the population. The regulatory assumptions that all crashes result in mortality are wrong, but there have been few studies on the possible repercussions of a collision or other stresses on animals. More research is required to determine the degree of harm and possible risk from crashes and other stresses to quantify the danger to individuals and groups appropriately.

2.8 Concerns About Using Wind Energy

It is expected that the capacity of wind energy plants will increase drastically. Figure 2.3 shows the expected rise of wind energy plants by 2030. Persistent concerns about the impact of this energy source on the environment hamper the widespread installation of marine and terrestrial wind farms. WREN's mission is to become the preeminent worldwide platform for sharing and disseminating accurate, up-to-date information on the latest, most reliable scientific research and methodologies for evaluating and keeping tabs on the environmental impacts of wind energy expansion. White papers developed by WREN will zero down on and further our knowledge of some important challenges facing the international wind sector [23].

Despite the global benefits of wind energy in reducing carbon dioxide emissions to mitigate climate change, opponents often object to the costs associated with ecosystem services and biodiversity through land/seascape

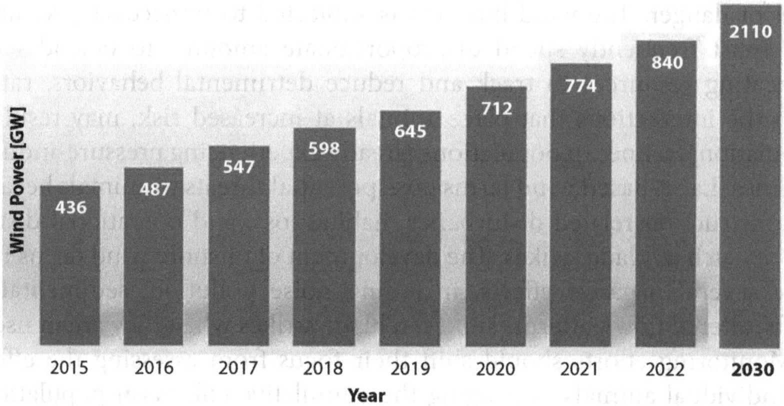

Figure 2.3 Expected cumulative capacity of wind energy plants by 2030. Reproduced from [16] under Creative Commons CC-BY license.

changes at smaller spatiotemporal scales. There is a gap between public (climate) benefits on the global scale and private (environmental) expenses on the local scale, and progress is hindered. Resolving "green *versus* green" conflicts for and against wind energy expansion requires planning and decision-making processes at intermediate geographical and temporal levels.

The approval procedure typically addresses environmental concerns from renewable energy production using the hierarchy of mitigation measures [23]. Active management (AM) has been advocated to actively apply the mitigation hierarchy throughout the life cycle of a wind energy project. Using iterative decision-making, AM may help plan and manage environmental hazards related to specific wind energy projects. The feasibility of using AM in wind energy projects to lessen negative environmental effects has been evaluated [18], along with the necessary monitoring at appropriate geographical scales. Wildlife within the footprint of individual wind energy facilities is a specific point of contention in environmental controversies. It is unknown, however, how these impacts may affect the species at greater regional ranges.

The standard practice has not yet provided conclusive criteria for shifting from emphasizing personal effects to focusing on community impacts, where the former indicates empirically derived quantities and the latter refers to societally acceptable thresholds. It is also difficult to generalize from local direct and indirect effects to population-wide demographic consequences.

Looking at the worldwide literature to find ways to get objective impact thresholds at the right geographical scales and to explain the connections between local and population-level impacts is desirable. Since a community may be impacted by several wind energy plants, graduating from local consequences to population-level impacts requires agreement across locations. How to define cumulative and transboundary effects and at what geographical scale they should be researched have been the subject of heated dispute. All over the world, there has been an increasing need for standardized approaches to integrating concerns about the cumulative impacts of wind energy into future studies and tracking practices. Biological diversity spans nations. Therefore, these norms must be implemented globally.

2.9 Conclusion

There is a need for a risk-based strategy to advise authorities and operators when approving a wind energy project due to the complexity of the problems associated with licensing and operational decisions. Associated risk estimation approaches may be used to examine and assess the environmental effects of wind energy's rapid expansion.

This information may then inform (adaptive) management decisions that aim to lower threats to endangered species populations, ecosystem health, and the likelihood of cumulative consequences. Wind energy and environmental groups might benefit from guidance on using risk-based methods for climate change mitigation, environmental conservation, and specific monitoring and mitigation initiatives.

The need for AM and the degree to which consequences must be handled at particular wind energy projects are grounded in "green vs. green" considerations and regional decision-making. Understanding the global effects on animals and the accompanying mitigation strategies requires the ability to extrapolate from the level of the individual to that of the population and then that of the biome. As a result, this knowledge might improve cumulative impact evaluations from the regional to the global level and encourage cross-border collaboration between wind power initiatives and their stakeholders. If regional decision-makers can access data on accumulating and transboundary repercussions, they may review the trade-offs and compromises between reducing global climate emissions and protecting local environments.

Trade-offs between climate change mitigation and neighborhood environmental effects, accumulated impacts of wind energy projects, and the difficulties in distinguishing population-level influences from effects on individuals can all be investigated using multidisciplinary specifications established and validated within a risk-based management framework. The investigations should combine a wide variety of research findings and perspectives on important holistic topics into a coherent whole. Finding a way to strike a balance between these concerns would enable technologists and policy makers to support the expansion of wind energy while limiting its negative effects on the environment.

References

1. A.K. Arya, A critical review on optimization parameters and techniques for gas pipeline operation profitability, *J. Pet. Explor. Prod. Technol.*, 12, 2022. 3033–3057. https://doi.org/10.1007/s13202-022-01490-5.
2. A.K. Arya, Application and Challenges of "Blockchain Technology" in the Oil and Gas Industry, in: *Blockchain Technology*, 2022: pp. 181–202. https://doi.org/10.1201/9781003138082-11.
3. A.K. Arya, S. Honwad, Modeling, simulation, and optimization of a high--pressure cross-country natural gas pipeline: Application of an ant colony optimization technique, *J. Pipeline Syst. Eng. Prac.*, 7, 2016. https://doi.org/10.1061/(ASCE)PS.1949-1204.0000206.
4. A.K. Thakur, A.K. Arya, P. Sharma, Prediction and mitigation of AC interference on the pipeline system, *Corros. Rev.*, (2022). https://doi.org/10.1515/corrrev-2021-0061.
5. A.K. Arya, R. Katiyar, P. Senthil Kumar, A. Kapoor, D.B. Pal, G. Rangasamy, A multi-objective model for optimizing hydrogen injected-high pressure natural gas pipeline networks, *Int. J. Hydrog. Energy*, 2023. https://doi.org/10.1016/j.ijhydene.2023.04.133.
6. OECD - Electricity exports by destination, IEA Electricity Information Statistics. 2012. https://doi.org/10.1787/data-00458-en.
7. A.K. Thakur, A.K. Arya, P. Sharma, The science of alternating current-induced corrosion: a review of literature on pipeline corrosion induced due to high-voltage alternating current transmission pipelines, *Corros. Rev.*, 38, 2020. https://doi.org/10.1515/corrrev-2020-0044.
8. S.S. Gupta, A.K. Arya, P. Vijay, Designing a model for optimization of maintenance and inspection efforts against third party damage to cross country Pipelines in India, Int. J.ournal Inno. *Tech. Exp. Eng.*, 8, 2019. https://doi.org/10.35940/ijitee.L3209.1081219.
9. A.K. Arya, S. Honwad, Optimal Operation of a Multi Source Multi Delivery Natural Gas Transmission Pipeline Network, *Chem. Prod. Process. Model.*, 13, 2018.1–17. https://doi.org/10.1515/cppm-2017-0046.
10. M.R. Islam, S. Mekhilef, R. Saidur, Progress and recent trends of wind energy technology, *Renew. Sust. Energ.*, 21, 2013. 456–468. https://doi.org/10.1016/j.rser.2013.01.007.

11. A.K. Arya, S. Honwad, Multiobjective optimization of a gas pipeline network: an ant colony approach, *J. Pet. Explor. Prod. Technol.*, 8, 2017 1389–1400. https://doi.org/10.1007/s13202-017-0410-7.
12. A.K. Arya, Optimal operation of a multi-distribution natural gas pipeline grid: an ant colony approach, *J. Pet. Explor. Prod. Technol.*, 11, 2021. 3859–3878. https://doi.org/10.1007/s13202-021-01266-3.
13. C.Q. Doe, Neural stem cells: balancing self-renewal with differentiation, Development. 135 (2008) 1575–1587. https://doi.org/10.1242/dev.014977.
14. Wind Electricity, No Title, IEA Electricity Information Statistics, 2022.
15. U. Singh, M. Rizwan, H. Malik, Wind Energy Scenario, Success and Initiatives towards, *Energies,* 15, 2022. 2291.
16. A.S. Darwish, R. Al-Dabbagh, Wind energy state of the art: present and future technology advancements, *J. Renew. Sustain. Energy*, 5, 2020. 7. https://doi.org/10.1051/rees/2020003.
17. S. Wang, S. Wang, Impacts of wind energy on environment: A review, *J. Renew. Sustain. Energy,* 49, 2015. 437–443. https://doi.org/10.1016/j.rser.2015.04.137.
18. A. Copping, L. Hanna, J. Whiting, S. Geerlofs, M. Grear, K. Blake, A. Coffey, M. Massaua, J. Brown-Saracino, H. Battey, Environmental Effects of Marine Energy Development Around the World. Annex IV Final Report, Office of Scientific and Technical Information (OSTI), 2013. https://doi.org/10.2172/1219898.
19. W. Musial, B. Ram, Large-scale Offshore Wind Power in the United States. Assessment of Opportunities and Barriers, Office of Scientific and Technical Information (OSTI), 2010. https://doi.org/10.2172/1219151.
20. A. Copping, L. Hanna, J. Whiting, Sharing Information on Environmental Effects of Wind Energy Development: WREN Hub, Wind Energy and Wildlife Interactions. 2017. 277–289. https://doi.org/10.1007/978-3-319-51272-3_16.
21. I.G. Simmons, Book review essays: Holling, C. S., editor, 1978: Adaptive environmental assessment and management. *Int. Inst. Appl. Syst. Anal.*, 3. Chichester: John Wiley. xiii+377 pp. £7.75, Progress in Human Geography. 3, 1979 631–632. https://doi.org/10.1177/030913257900300423.
22. J. Köppel, M. Dahmen, J. Helfrich, E. Schuster, L. Bulling, Cautious but Committed: Moving Toward Adaptive Planning and Operation Strategies for Renewable Energy's Wildlife Implications, *Environ Manage.*, 54,2014 744–755. https://doi.org/10.1007/s00267-014-0333-8.
23. R. May, A.B. Gill, J. Köppel, R.H.W. Langston, M. Reichenbach, M. Scheidat, S. Smallwood, C.C. Voigt, O. Hüppop, M. Portman, Future Research Directions to Reconcile Wind Turbine–Wildlife Interactions, Wind Energy and Wildlife Interactions, 2017 255–276. https://doi.org/10.1007/978-3-319-51272-3_15.
24. J.M. Kiesecker, H.E. Copeland, B.A. McKenney, A. Pocewicz, K.E. Doherty, Energy by Design: Making Mitigation Work for Conservation and Development, Energy Development and Wildlife Conservation in Western North America, 2011 159–181. https://doi.org/10.5822/978-1-61091-022-4_9.

25. I. Rushworth, S. Krüger, Wind farms threaten southern Africa's cliff-nesting vultures, Ostrich. 85, 2014. 13–23. https://doi.org/10.2989/00306525.2014.913211.
26. E.L. Dahl, K. Bevanger, T. Nygård, E. Røskaft, B.G. Stokke, Reduced breeding success in white-tailed eagles at Smøla windfarm, western Norway, is caused by mortality and displacement, *Biol. Conserv.*, 145, 2012 79–85. https://doi.org/10.1016/j.biocon.2011.10.012.
27. J. Bellebaum, F. Korner-Nievergelt, T. Dürr, U. Mammen, Wind turbine fatalities pproach a level of concern in a raptor population, *J. Nat. Conserv.*, 21, 2013. 394–400. https://doi.org/10.1016/j.jnc.2013.06.001.
28. J.C. Garvin, C.S. Jennelle, D. Drake, S.M. Grodsky, Response of raptors to a windfarm, *J. Appl. Ecol.*, 48, 2010. 199–209. https://doi.org/10.1111/j.1365-2664.2010.01912.x.
29. J.A. Shaffer, D.A. Buhl, Effects of wind-energy facilities on breeding grassland bird distributions, *Biol. Conserv.*, 30, 2015 59–71. https://doi.org/10.1111/cobi.12569.
30. A. Sansom, J.W. Pearce-Higgins, D.J.T. Douglas, Negative impact of wind energy development on a breeding shorebird assessed with a BACI study design, *Ibis.* 158, 2016. 541–555. https://doi.org/10.1111/ibi.12364.
31. R. Brabant, N. Vanermen, E.W.M. Stienen, S. Degraer, Towards a cumulative collision risk assessment of local and migrating birds in North Sea offshore wind farms, *Hydrobiologia.*, 756, 2015 63–74. https://doi.org/10.1007/s10750-015-2224-2.
32. J.E. Diffendorfer, J.A. Beston, M.D. Merrill, J.C. Stanton, M.D. Corum, S.R. Loss, W.E. Thogmartin, D.H. Johnson, R.A. Erickson, K.W. Heist, Preliminary methodology to assess the national and regional impact of U.S. wind energy development on birds and bats, *Scientific Investigations Report*, 2015. https://doi.org/10.3133/sir20155066.
33. A. Sanz-Aguilar, J.A. Sánchez-Zapata, M. Carrete, J.R. Benítez, E. Ávila, R. Arenas, J.A. Donázar, Action on multiple fronts, illegal poisoning and wind farm planning, is required to reverse the decline of the Egyptian vulture in southern Spain, *Biol. Conserv.*, 187, 2015. 10–18. https://doi.org/10.1016/j.biocon.2015.03.029.
34. R.E. Green, R.H.W. Langston, A. McCluskie, R. Sutherland, J.D. Wilson, Lack of sound science in assessing wind farm impacts on seabirds, *J. Appl. Ecol.*, 53, 2016 1635–1641. https://doi.org/10.1111/1365-2664.12731.
35. J. Nabe-Nielsen, R.M. Sibly, J. Tougaard, J. Teilmann, S. Sveegaard, Effects of noise and by-catch on a Danish harbour porpoise population, *Ecol. Modell.*, 272, 2014. 242–251. https://doi.org/10.1016/j.ecolmodel.2013.09.025.
36. B. Ram, Integrated Risk Framework for Gigawatt-Scale Deployments of Renewable Energy: The U.S. Wind Energy Case; October 2009, Office of Scientific and Technical Information (OSTI), 2010. https://doi.org/10.2172/977302.

3

Important Issues and Future Opportunities for Huge Wind Turbines

Anjali Tripathi[1], Vinay Kumar Pandey[2], Shivangi Srivastava[3], Rashi Mishra[4] and Ashish[5]*

[1]Department of Biotechnology, Sharda School of Engineering and Technology, Sharda University, Greater Noida, India
[2]Division of Research & Innovation, School of Applied and Life Sciences, Uttaranchal University, Dehradun, Uttarakhand, India
[3]Department of Food Technology, Harcourt Butler Technical University, Nawabganj, Kanpur, Uttar Pradesh, India
[4]Department of Biotechnology, Axis Institute of Higher Education, Kanpur, Uttar Pradesh, India
[5]Department of Bioengineering, Integral University, Lucknow, Uttar Pradesh, India

Abstract

Huge wind turbines are built on a big scale and are intended to produce a lot of electricity. They are frequently grouped together in wind farms and are normally significantly larger than the small-scale wind turbines utilized for residential and commercial uses. They often have capacities from hundreds of kilowatts to several megawatts, which is far higher than those of residential and commercial turbines. Huge wind turbines are crucial because they employ the wind's energy to produce clean, sustainable, and renewable energy. As an alternative energy alteration for both commercial and domestic use, the use of wind turbines are increasing in popularity. Large wind turbines can generate enough electricity to power entire cities or small countries by utilizing this unrestricted and free natural resource. Large and small wind turbines are becoming more and more popular as alternative energy sources. As part of their initiatives for renewable energy, a lot of nations and regions are investing in wind energy. It is possible for large wind turbines to produce enormous amounts of electricity enough to power entire cities or even small nations. Significant problems with huge wind turbines and current challenges are explored in this chapter.

Keywords: Wind turbines, sustainability, environment, renewable energy

**Corresponding author*: ashish.itbhu13@gmail.com

Inamuddin, Tariq Altalhi and Mohammad Luqman (eds.) Wind Energy Storage and Conversion: From Basics to Utilities, (33–62) © 2024 Scrivener Publishing LLC

3.1 Introduction

Fossil fuel use is widely acknowledged as a major contributor to global warming. As a result, energy experts, business executives, and lawmakers have switched their attention more and more towards renewable energy sources. Geothermal, wind, and biomass technologies have all made great strides and are now more competitive than ever. According to the European Renewable Energy Council, by 2040, renewable energy sources might supply nearly half of the world's energy requirements. The efficiency and advancement of renewable energy sources are anticipated to significantly rise before 2050, resulting in considerable reductions in carbon dioxide emissions. National governments, urban planners, and others who want to switch to cleaner energy sources, reduce CO_2 emissions, promote economic growth, and generate jobs have taken notice of wind power's spectacular recent growth. The most recent Global Wind Report indicates that by the end of 2013, there were 318,105 global wind units installed worldwide. It is crucial to recognize that the development of wind energy also poses some environmental difficulties. Significant concerns have been voiced within the wind energy industry over the environmental effects of wind energy, particularly in developing nations and ecologically vulnerable locations. To strike a balance between energy production and environmental sustainability, it is imperative to solve these issues.

3.1.1 Visual Impact

Wind turbines' aesthetic impact relates to how they seem and the way they may change a landscape's scenic appeal. The presence of wind turbines can be aesthetically striking, especially when they are grouped together in wind farms. Wind turbines are frequently towering buildings with huge revolving blades. In Germany and many other nations, a notable quantity of wind turbines has been constructed to make a shift to renewable energy. Several research have been conducted to determine the underlying causes of wind turbines subjectively perceived as visual influence on the aesthetic value of landscapes. However, no research has ever been done on how much moral opinions about wind turbines affect how attractive they are. Researchers investigated the way suggested judgements and moral associations of various huge pole-like constructions, such as incinerator plant chimneys, wind turbines, and high-frequency communication towers, affected claims made concerning the way these structures affected the aesthetic value of landscapes to tackle this problem. The study reveals such different pole-like constructions are considered to have varying degrees of visual quality and scenic beauty impairment, even though they appear to differ a little up close. This may be explained by the structure's markedly varied moral connections, which lead to the judgment that a communication tower is a medium impairment while an incineration chimney is assessed to be by far the largest. Both advocates and detractors of wind energy can agree on this association. However, these groups' opinions of wind turbines diverge significantly: supporters place much less weight on the negative effects than detractors. These results together strongly imply that the moral connotations of mast-like structures in landscapes have a major impact on scenic beauty judgments [1]. The landscape can change significantly in locations with dense populations of wind turbines, such as wind farms. The presence of these structures might change the typical views of broad fields, undulating hills, or seaside panoramas. The visual impact may be particularly obvious if the wind turbines are placed in aesthetically delicate places, such as scenic landscapes, important historical sites, or popular tourist destinations. Individuals have different subjective aesthetic opinions about wind turbines. Some individuals find the sleek, futuristic design of wind turbines to be appealing and see them as emblems of technical advancement and renewable energy. Others can find them distracting or aesthetically invasive to the area's natural or cultural character. When people or other stakeholders find the aesthetic impact of wind turbines to be objectionable, local communities may become opposed. Changing beloved landscapes, implications on property prices, or alleged detrimental effects on tourism and recreation are just a few possible causes for concern. Legal difficulties or delays in project permits may occasionally result from objections to wind turbines based on their visual impact. These issues need to be addressed by developers and planners through thorough visual impact analyses and community participation procedures. There are mitigation strategies that can be used, such as picking suitable turbine designs, placing them in strategic locations, or using landscaping strategies to blend turbines into the surrounding environment [2]. A constant issue is striking a balance between the requirement to produce renewable energy and maintaining the aesthetic value of landscapes. Site selection must be carefully thought out, local communities must be consulted, and aesthetic standards and planning requirements must be followed. These factors can be considered to lessen the aesthetic impact of wind turbines and increase public support for renewable energy initiatives [3].

3.1.2 Noise

Globally, there is utilization of wind energy in the form of clean source of energy. However, the low-frequency noise (LFN) produced by wind turbines ranges from 20 to 200 Hz, and it can either be heard or not [4]. Aerodynamic noise and mechanical noise are the two primary categories of LFN produced by wind turbines. The blades of turbines revolve sluggishly or are "placed" or immobile when the speed of the wind is lower than the cut-in speed, making very little noise [4]. When a wind turbine operates between speeds of roughly 4 and 30 m/s, as measured at the hub position, its sound power level continually rises. The main source of noise produced by wind turbines is aerodynamic noise, which is brought on by air moving over the blades. Blade tips, turbulence in the entering wind, impulsive noise, contact between the blades and the tower, and noise at the trailing edge of the blades are some of the sources of aerodynamic noise [5]. Higher amounts of aerodynamic noise arise from the wind turbine's rotating speed increasing along with the wind speed. Low-frequency noise (LFN) from wind turbines has been linked to symptoms like annoyance, stress, insomnia, migraines, tinnitus, exhaustion, anxiety, and possibly hearing loss, according to a number of studies. Additionally, vibroacoustic sickness, chronic weariness, and attention deficit disorder have all been linked to LFN, which raises the risk of seizures, heart issues, and coronary artery disease [6].

The harmful effects of low-frequency noise (LFN) are felt by more people as noise levels rise. People who live close to wind farms frequently are exposed to LFN. Using LFN propagation models, it is possible to gauge or estimate how much exposure residents have to turbine LFN. The paperwork provided by turbine manufacturers and other sources usually lacks important details regarding the sound power level of wind turbines (LW,A). Only the complete audio spectrum (20–20 kHz) or certain frequency ranges, such as 63–8 and 50–10 kHz, are covered by some sources' LW,A values. Unfortunately, turbine manufacturers hardly ever publish LW,A values at 20–200 Hz. This work attempted to fill up this gap by approximating the LW,A parameter, which was then applied to the estimation of LFN exposure from wind turbines. This information is important to determine the level of LFN exposure experienced by surrounding residents and determine any potential health impacts of turbine LFN. Determining the citizens' exposure to LFN from turbines is crucial before studying the effects of LFN on health. Although it is possible to directly quantify their LFN exposure, the labor and resource requirements are challenging. Rain, wind speed, and wind direction are among the environmental and meteorological variables that might affect noise transmission [7]. It was found in a study by Ma and Nystuen [8] that even mild rainfall results in higher sound levels than wind. They did add that rain had no effect on the dispersion of sound at natural frequencies. Results from earlier research on the impact of precipitation on noise propagation have been ambiguous. This study's [9] goal is to ascertain how, under typical climatic conditions, rainfall especially influences the propagation of low-frequency noise (LFN) from wind turbines. It is important to keep in mind that the multiplicity of LFN sources in cities may reduce the noise's ability to travel through the air. In urban areas, LFN is frequently present as background noise and is produced by a variety of man-made sources, including air-conditioning equipment, industrial machinery, pumps, and vehicles on land, in the air, on trains, and in the water [10]. The sources may be located far away because LFN can travel long distances. Taiwan is a little island with a large population. In Taiwan, wind farms have been built close to residential areas, directly affecting the daily life of the locals. An estimate of the residents' LFN exposure (LAeq) could be made using an ISO 9613-2-compliant low-frequency propagation model. However, the manufacturers failed to supply the necessary parameter for this model, which was the sound power level of wind turbines (LW,A (dB)) at 20–200 Hz [11].

3.1.3 Wildlife

Similar to other sources of energy, wind energy, too, has the possibility of damaging the environment through lowering, degrading, or destroying habitats for fish, animals, and plants. Besides these, flying animals like birds and bats may be put at risk by turbine blade rotation. Impact decrease, siting, and regulatory issues have been the top priorities of the wind business due to the possibility of wind power having a drastic effect on the fauna and the fact that those challenges can either detain or avert wind utilization in elevated standard of wind assessed locations. To solve those matters and promote an eco-friendly sustainable growth of wind energy in the USA, the World Trade Organization sponsors initiatives that try to shed light upon the adverse impacts of wind on land on air and offshore [12]. Wind Energy Technologies Office (WETO) conducts projects for compiling and dispersing equally reviewed studies based on the environmental effects by incorporated detailed hubs such as Tethys. An agency additionally raises funds for scientific research which sanctions the formation of affordable technology to lessen the effects on animals at onshore and offshore wind farms. To guarantee that taxpayer funds are spent properly to address environmental concerns interconnected to the utilization of wind in the United

States of America, WETO functions in promoting the interorganizational partnership on the impacts of wind energy and sedentary studies [13].

Over the last decade, the NWCC has conveyed its emphasis on inscribing or spreading top-notch knowledge about environmental impacts and solutions. The wind office of the Department of Energy announced $2 million in ecological experiment allowed in May 2009 with the aim to lessen the risks concerning wind power development to critical species and habitats [14]. In an investigation made in 2013, scientists from the NWCC's Grassland Community Collaborative and the Kansas State University demonstrated that the wind expansion in Kansas showed no effect on residents and the replication of larger steppe fowl. Earlier founded in 2003, the National Renewable Energy Laboratory of DOE had funded several research projects involving bats and wind energy cooperative, including investigations into the effects on modifying the cut-in speed of wind turbines (the least speed of wind at which the power production of wind turbine begins) and usage of supersonic auditory deterrent to lessen the impacts on bats of wind turbines. WETO finances development and research initiatives, too, which improves the scientific willingness for minimization of bat impacts and reductive solutions between fierce financial chances [15].

The Department of Energy is providing funding to Frontier Wind, Bat Conservation International, Texas Christian University, the University of Massachusetts, and General Electric to experiment, analyze, and check prevention technologies for the impact of bats non-commercially. These technologies would give wind facility owners–operators and regulators both affordable and practical instruments to lessen the impacts on bats [16]. The National Wind Coordinating Collaborative's seminars on the Findings and Standards of evolving technical innovations for Deterrence at Wind Facilities and Bat detection present testing results and project updates until March 2018. In 2016, WETO selected approximately six groups to task on enhancing strategies which would protect eagles which might allocate the airway with wind turbines. Nearly $3 million was distributed to the six teams for ground-breaking, essential research, and development initiatives involving eagle impact mitigation technology. The investigation that was supported by this award would provide the owners and managers of wind farms with affordable and useful methods to minimize possible impacts on eagles. By acknowledging environmental and sedentary issues, this significant study advances the Energy Department's initiatives to assist the growth of renewable sources of energy while simultaneously guaranteeing animal cohabitation. If the project is successful, it will protect biodiversity and provide the renewable energy sector with fresh capabilities to ease administrative and financial worries [17]. WETO is an advocate for the study of how biological processes interact with offshore wind farms. With the help of this financing, scientists are collecting vital information on offshore bird, bat behavior, marine life, and various aspects which affect the offshore wind turbine placements in the United States–for instance, to create an in-depth understanding of ecosystem in a Mid-Atlantic Wind Energy field which would help with enabling ecological conformance for the Biodiversity Research Institute, a diverse group of collaborators and offshore wind projects completed the major environmental study that is ever carried out in the Mid-Atlantic.

WETO, along with other federal agencies, focuses and runs to make suggestions which will aid landowners in adhering to legal, managerial, and regulatory requirements for public safety, national security, and the preservation of wildlife [19].

3.1.4 Intermittent Energy Generation

The maturation of renewable sources of energy (such as geothermal energy, wind energy, biomass energy, and solar energy) has emerged as main agreement and decisive plans for nations globally as challenges of environmental pollution and energy crisis become more urgent [20]. Wind power has now vigorously set up its own self as a conventional alternative of humanity among all renewable energies. There are three key factors that contribute to wind energy usage globally. First, there has been an endless supply for wind energy. According to statistical data, there are more than 1 trillion kilowatts of onshore wind resources available. About 840 PW·h of potential power from wind is thought to exist [21]. Second, wind energy has significant environmental advantages, including reduction in air pollution and almost minimal water use. The decrease in carbon dioxide emissions is wind energy's biggest environmental advantage [22]. The entire number of wind farms in operation reduced carbon dioxide emissions by around 372 million tonnes in 2013. Finally, the cost of wind energy is decreasing as technology gets more advanced.

Several policies and support systems have been put in place to advance the growth of wind power. These programs include fixed premium systems, renewable portfolio standards, investment subsidies, and tax incentives. Due to the benefits and favorable legislative environments, wind power has grown rather quickly over the past few decades on a global scale. By the end of 2015, the globally progressive setup potential acquired 432,419 MW, which is around 25 times more than the setup potential in year 2000 [23]. The installed wind potential is predicted for acquiring 2000 Global Wind by the end of year 2030, providing roughly 16.7–18.8% of the total power produced globally. In 2015, wind energy produced 42.1% of the electricity in Denmark. By 2050, the government of Denmark wants to have 50% of its electricity coming from wind [24]. In recent decades, wind energy development in China has experienced an exponential growth. Despite a decline in 2013, the cumulative installed wind energy capacity in the country reached 91.42 global wind units, positioning China as the global leader in wind energy installations [25]. Then, wind energy will meet 17% of the country's electricity needs and establish itself as a significant energy source. Wind power has advanced quickly, but integration and consumption are not encouraging [26]. Wind energy is regularly curtailed. The average number of hours that wind power was used in 2015 was 1,728, which is the lowest number over the previous 6 years. The rate of wind power curtailment over the same time is 17% (the turbines are fictitiously turned away notwithstanding the normal flow of wind) [27, 28].

The wind's varied speed, which varies both spatially and temporally, is what drives wind turbines. Consequently, the electricity produced by turbines is typically very sporadic. Particularly, wind energy might not be readily accessible [29–31]. As with thermal, nuclear, and hydroelectric plants, wind generation cannot be scheduled or managed [32]. Large-scale penetration of wind power would consequently have an effect on the operational safety and steadiness of the power system, which will raise costs. The largest obstacle with usage of wind energy in the form of primary energy source is now its erratic nature. In Figure 3.1, a computational simulation of wind power generation is explained. System design, analysis, and optimization for wind power generation all heavily rely on computational simulation. It

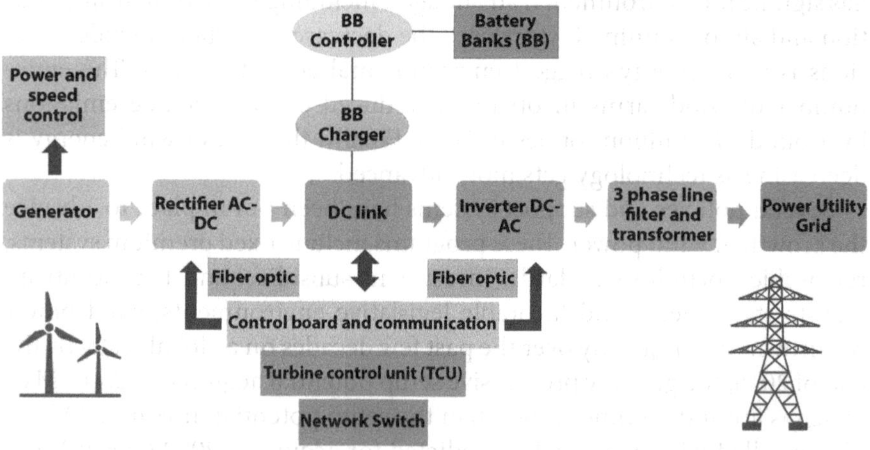

Figure 3.1 Computational simulation of wind power generation.

enables engineers and scientists to evaluate the efficacy and efficiency of wind turbines, calculate power production, analyze the effects of environmental conditions, and optimize the design of the entire system [68].

3.2 Worldwide Wind Energy Forecast

The devastating effect of increased global warming has led the world towards the advancement of energy sources. With technological advancements, the world is opting for low-carbon or carbon-free emission products, sustainable as well as eco-friendly renewable energy. Today as the world is moving towards the utilization of renewable energy sources, wind energy has reached an epitome as the best renewable energy source. The adoption of wind power as a widely used renewable energy can be efficient, effective, appropriate, and practically able to lessen the fossil fuels' utilization. The sector concerned with wind power usage is having remarkable growth and has golden future opportunities. According to the data, in the 21st century, the use and implementation of wind production has grown approximately two times every 3 years. The scenarios faced by each country are discussed below.

3.2.1 Canada

In the upcoming supply of energy globally, the importance of renewable wind source of energy has been analyzed statistically in Canada. A survey was conducted to study the capacity factor, installation, generation, and awareness regarding this implementation, supplied upon the levelized cost of energy (LCOE). This survey was made to present an inclusive fundamental for a further approach of the country towards opting wind energy and its potential development in the electric supply of Canada to the upcoming 2040 by applying different prognostic methods, involving historical compound annual growth rate (CAGR). CAGR of 5%, 10%, or 15% and linear forecasting was appointed based on the assumed terms and policy intervention at the territorial or provincial level. Additionally, the impact of rising installation of wind energy potential in Canada on the LCOE is inspected through making use of a linear forecasting model alongwith several reduced prices for costs. The outcomes depict the deduction in the LCOE based upon a higher installation of wind power capacity. In conclusion, Canada experiences a high use of wind energy power and has a propitious installation of wind energy potentials, specifically in St. Lawrence, whether integrated with the other energy source or as standalone source of energy, with the storage potential of energy to converge the upcoming generation's perspective of balancing low-carbon footsteps with the electricity requirements [33].

3.2.2 Russia

The Russian energy industries are genuinely acting in accordance with the standard conventional growing system. The accessibility of water resources and large hydrocarbon stock, e.g., natural gases, coal, and oil, has transformed into a notable barrier obstructing the development of efficient and safe substitute sources of energy, yet wind energy is comparatively a small contributor in the electricity generation mix in Russia, but the country has significant untapped potential for wind power development. The government of Russia has set an objective of generating 4.5% of the country's electricity from renewable sources by 2024, and wind power is thought of being involved in the significant role in acquiring the same goal. According to the Global Wind Energy Council, Russia had a setting-up wind energy potential of 0.6 GW at the end of 2020, which accounted for only 0.1% of the country's entire electricity generation. The Russian government has a proposal for the increment in the country's wind energy potential to 3.2 GW by year 2024 and up to 11 GW by 2035 [34]. These plans are supported by favorable wind conditions in certain regions of the country, particularly in the eastern and southern parts of Russia. To support the growth of wind power in Russia, the government has introduced various incentives and support mechanisms, including feed-in taxes and tax exemptions. The maturation of wind power energy in Russia is still facing some challenges, including the relatively high cost of wind power and insufficiency of infrastructure compared to other energy sources. An outlook for wind power in Russia is promising, and the country has the possibility to become a main player in the global wind power market in the coming years. The most challenging issue faced by Russia with respect to energy security involves raised energy share in the energy balance, as well as global fuel, and raised international attempts for the execution of climatic terms and policy and encourage the transformation into a green economy [33].

3.2.3 India

India has been quickly expanding its capacity for renewable energy, with a particular emphasis on wind energy. India is projected to be the fourth-largest wind energy market in the world by 2021, with the capacity to build 38

Global Wind power facilities. By 2030, the government wants to reach 450 GW of renewable energy capacity, 140 GW of which will come from wind power. India has put into effect regulations that encourage the growth of wind energy projects, such as the KUSUM Outline and the National Wind–Solar Hybrid Policy. India has made great strides in the forecasting of wind energy. All grid-connected wind energy generators must use the Wind Energy Forecasting and Scheduling (WEFS) system, which was established by the Ministry of New and Renewable Energy (MNRE). This system forecasts wind energy production using weather forecasting models and offers scheduling data to guarantee seamless grid integration. India is expected to develop 20 GW of wind energy capacity between 2021 and 2025, according to the Global Wind Energy Council (GWEC), demanding precise wind energy forecasting for grid stability. With a goal of 5 GW capacity by 2022, the paper also underlines India's potential for offshore wind energy. The fluctuation in wind speed and direction, which is impacted by factors like seasons, geography, and land use, is one of the major issues in wind energy forecasting. The MNRE has developed wind monitoring stations across the nation to solve this issue and provide real-time data for forecast models. In order to improve the precision of wind energy forecasting, the MNRE is also investigating the use of machine learning methods. The government's goals and advances in wind energy forecasting technology are anticipated to significantly boost growth in the renewable energy sector in the upcoming years. India's position as the world's leading producer of wind energy is expected to promote the country's transition to a low-carbon energy system [33].

3.2.4 United States of America

The United States is one of the world's biggest wind energy markets, with about 119 GW of installed capacity as of 2021. Wind energy is essential to the country's ongoing energy portfolio because the government has set goals for the use of renewable energy sources and the lowering of emissions. In recent years, the United States has made significant progress in wind energy forecasting. The National Renewable Energy Laboratory (NREL) has developed advanced forecasting models that utilize historical data on wind energy generation as well as weather data in order to precisely estimate wind energy output. These models take into account a variety of factors, such as wind direction, speed, humidity, temperature, and atmospheric pressure. The United States has also implemented policies like the Production Tax Credit (PTC) and the Investment Tax Credit (ITC), which provide financial incentives to stimulate the expansion of wind energy facilities. These rules are intended to promote the growth of wind energy projects all around the country [34]. Several US states have implemented Renewable Portfolio Standards (RPS), which mandate that businesses obtain a specific proportion of their energy from renewable sources like wind energy. These governments have set RPS rules and objectives for renewable energy. According to the American Wind Energy Association (AWEA), the nation's installed wind energy capacity will rise by 14.2 GW in 2020, reaching a total of 119 GW. This analysis also highlights the potential for offshore wind energy exploration in the country, where a number of sizable offshore wind projects are now in the development stages. Wind energy is anticipated to have a significant impact on the low-carbon energy system in the United States. The government encourages the growth of wind turbines, which is anticipated to occur in the next years, by using forecasting tools [33].

3.2.5 China

With a total installed capacity expected to reach 281 GW in 2021, China is now the largest wind energy market in the world. China's renewable energy portfolio includes a sizable amount of wind energy, and the country's government has high goals for both renewable energy production and carbon reduction. In recent years, China has achieved notable advancements in wind energy forecasting. The National Meteorological Information Centre (NMIC), which is in charge of forecasting wind energy, has created sophisticated models that use previous data on wind energy generation and weather data to precisely estimate wind energy output. These forecasting models take into account variables including wind speed, direction, humidity, temperature, and atmospheric pressure. The NMIC uses a nationwide network of meteorological stations to collect real-time weather data, improving the precision of the forecast models. The research also emphasises China's potential for developing offshore wind energy, where a number of sizable offshore projects are now in the planning stages. The fluctuation in wind speed and direction, which is impacted by elements like terrain, climate, and atmospheric conditions, is one of the major issues in China's wind energy forecasting. The National Renewable Energy Centre (NREC) has created regional wind forecasting models that take into account local weather patterns and terrain to address this issue. Through the application of machine learning algorithms and other cutting-edge approaches, the NREC is also actively seeking to increase the accuracy of wind energy forecasts [33].

3.2.6 Germany

German wind energy production is among the best in the world, with about 62 GW of installed capacity as of 2021. With the government setting goals for renewable energy generation and emissions reduction, wind energy is a key component of Germany's renewable energy portfolio. The German Meteorological Service (DWD) has created sophisticated forecasting models, which have allowed for considerable breakthroughs in wind energy forecasting in recent years. In order to effectively anticipate wind energy output, these models use weather data as well as historical wind energy generation data, taking into account elements like temperature, wind direction, wind speed, humidity, and atmospheric pressure. To encourage the development projects of wind energy, the nation has established a number of measures, including the Renewable Energy Act (EEG). The overall installed capacity of wind energy in Germany increased by 1.6 GW in 2020, reaching over 62 GW, according to a report by the German Wind Energy Association (BWE). However, due to legislative barriers and opposition from the general public, the development pace of the wind energy sector has recently slowed. The government has recently introduced measures to address these challenges, including new regulations for the distance between wind turbines and residential areas. To address this challenge, the DWD has developed regional wind forecasting models that consider local weather patterns and topography. The DWD is also working on improving the accuracy of wind energy forecasting using machine learning algorithms and other advanced techniques [38]. However, there are still challenges facing the wind energy sector, including regulatory barriers, public opposition, and variability in wind speed and direction. Governments, industry, and other stakeholders will need to work together to overcome these challenges and support the growth of the wind energy sector.

3.3 Increased Wind Penetrating Techniques

The installation of wind energy capacity has increased the future opportunities for setting up of wind turbines across the world. The techniques used are discussed below.

3.3.1 Energy Storage Systems

Energy storage systems could be utilized for the levelized fluctuations in the output of wind power and provide grid stability during periods of low wind. Pumped hydro storage, compressed air energy storage, and battery storage are some instances of storage systems for energy.

3.3.2 Advanced Forecasting Tools

Advanced forecasting tools, such as weather modeling and machine learning algorithms, can improve the precision for wind power prediction, which could help grid engineers to foresee amendments in output of wind power and make real-time adjustments to maintain grid stability [33].

3.3.3 Bucket Foundation

Bucket foundation was established for offshore gas and oil industries and is now used for the construction of wind turbines. Bucket foundation is basically used for supporting the construction of structures like wind turbines and gas or oil platforms. Bucket foundation comprises a cylindrically large steel bucket that is driven into the seabed with the use of a specialized pile driving technique. After the installation of bucket, it is filled with gravel or any other materials to provide additional stability and weight. The structure which is to be constructed or supported is then robust on the top of the bucket [36].

3.3.4 Advantages of Bucket Foundation

A bucket foundation provides a variety of advantages in contrast to other offshore technologies; firstly, it provides ease of installation. They are an efficient, simple technique that reduces the cost as well as time which is needed for offshore construction projects. Secondly, being highly stable, they also promote highly notable stability and load-bearing capacity, which makes them ideal for supporting heavy structures in challenging environments. Thirdly is versatility; they are highly resistant to lateral loads, such as those caused by strong ocean waves or currents. This is supported because the bucket behaves as a giant anchor, which provides a wide base for the structure to surpass these high forces. Additionally, the use of a steel bucket supplies excellent corrosion resistance, which is essential for offshore applications. This quality of being adaptive to every soil type increases flexibility, which makes them a popular choice for offshore projects which require complex or unique foundations. Fourthly is cost-effectiveness; they reduce the rate of cost and require less time to construct. In contrast to traditional pile foundations, they require less labor and less time and are relatively inexpensive to install and manufacture, making them an attractive option for offshore projects with tight budgets [35].

3.3.5 Limitations of Bucket Foundation

In spite of having various advantages, bucket foundations have few limitations. One of their main limitations is their ability to withstand lateral loads, such as those caused by wind and waves. To address this issue, engineer, designers often incorporate extra measures, such as pile anchors or additional weight, into the foundation design. In short, bucket foundations are a cost-effective, versatile solution for offshore projects which need stable and reliable foundations. Their ease of installation, flexibility, and customization options make them a popular choice for a wide range of offshore applications [33].

3.3.6 Monopile Foundation

A monopile foundation consists of a large steel tube driven into the seabed to provide structural support and anchorage for the turbine. The monopile foundation is designed to withstand the forces of the wind, waves, and currents that are present in offshore environments. The foundation is typically driven into the seabed using a pile driving system, which involves using a large hammer to pound the steel pole into the seabed. Once the monopile is in place, the turbine is lodged upon it. For offshore wind farms, monopile foundations are a popular choice as they are relatively easy and cost-effective to set up. They are also able to support the weight of large wind turbines and withstand the harsh offshore environment. However, they may not be suitable for all seabed types, and alternative foundation types may be necessary in certain locations [35].

3.3.7 Jacket Foundation

A jacket foundation is a type of offshore foundation used to support wind turbines in offshore wind farms. It consists of a steel lattice structure that is designed to withstand the forces of the wind, waves, and currents that are present in offshore environments. The jacket foundation is typically installed by firstly driving a set of piles into the seabed to provide support for the structure. The jacket is then assembled onshore and transported to the installation site, where it is lifted onto the piles and bolted in place. Once the jacket is secured to the seabed, the turbine is lodged upon the cap of it. Jacket foundations are a popular choice for offshore wind farms because they can support larger wind turbines than monopile foundations and are generally more stable in deeper waters. They are also able to withstand the harsh offshore environment, including extreme weather conditions and corrosive saltwater. However, jacket foundations are more complex and expensive to install than monopile foundations due to the additional fabrication and transportation required. They also require a larger installation vessel, which can limit their use in certain offshore locations. Despite these limitations, jacket foundations are still a popular choice for offshore wind farms, particularly in areas with deeper waters and larger turbines. Ongoing advancements in technology and fabrication techniques are also helping to reduce the costs associated with jacket foundation installations [33].

3.3.8 Floating Foundation

A floating foundation is a type of offshore foundation used to support wind turbines in deep water or areas where fixed foundations are not suitable. It consists of a buoyant platform that is moored to the seabed with the use of anchors or mooring lines, which maintain the position of the platform and its steadiness. The two major kinds of floating foundations are tension leg platforms (TLPs) and semi-submersible platforms. TLPs use tensioned cables to anchor the platform to the seabed, while semi-submersible platforms use ballast to submerge the lower portion of the platform and maintain its stability. Floating foundations are able to support wind turbines in deeper waters than fixed foundations and can be moved to different locations as needed [36]. They also have the potential to generate more electricity than fixed foundations, as they are not limited by water depth. However, floating foundations are generally more expensive to install and maintain than fixed foundations due to their more complex design and the need for regular maintenance of the mooring lines. They also require more space than fixed foundations, as the platform needs to be able to move with the wind and waves. Despite these challenges, floating foundations are a mandatory technology for the advancement of offshore wind power in deeper waters. Ongoing advancements in technology and manufacturing are helping to reduce the costs associated with floating foundation installations, making them a more viable option for offshore wind farms in the future [33].

3.3.9 Tripod Foundation

A tripod foundation is a type of offshore foundation used to support wind turbines in shallow water environments. It consists of three steel legs that are connected at the top by a circular steel structure, which serves as the main support for the wind turbine. The tripod foundation is installed by firstly driving the legs into the seabed using a pile driving system. The circular structure is then placed on top of the legs and secured in place. Tripod foundations are a popular choice for offshore wind farms with shallow water, as they are relatively easy and cost-effective to employ. They are also able to support the weight of large wind turbines and withstand the forces of the wind, waves, and currents that are present in offshore environments. However, tripod foundations may not be suitable for all seabed types, and alternative foundation types may be necessary in certain locations. They are also not suitable for deeper waters, where other types of floating or fixed foundations are used. These foundation system technologies are continually being developed and improved to expand productivity and decrease costs. A choice for the foundation systems of wind farms is based upon various factors like water depth, seabed conditions, and turbine size [33–37].

3.4 India's Perspective for Wind Energy

India is among the rapidly expanding and world's largest economies, with high demand for energy to power its industries and population. However, its heavy reliance on fossil fuels has led to significant environmental and economic challenges, including air pollution, greenhouse gas emissions, and energy security concerns [47]. India has been exploring alternative forms of energy, including wind power, converging with its growing energy requirements while mitigating the adverse effects of fossil fuels. Wind energy proved to emerge as the most favorable alternative energy form for India, given the large coastline and vast land resources. In Figure 3.2, a wind turbine system is given, which helps to clearly understand the concept as to how it works. India has a cumulative wind power potential of about 40.57 Global Wind as of March 2021—all over the world after China, USA, and Germany—making it the fourth-largest wind energy producer. The nation has set up an objective of acquiring 60 Global Wind of wind power potential by year 2022, which will require an addition of around 20 GW in the next 2 years. In India, the capability of wind energy is extensive, given its favorable geographical and meteorological conditions. The country has a coastline of around 7,500 km, which provides significant opportunities for offshore wind power generation. According to the National Institute of Wind Energy (NIWE), India has offshore wind capacity of about 70 GW, which is more than twice its current wind energy capacity. In addition, India has vast land resources, particularly in the western and southern regions, where wind speeds are high and consistent [48].

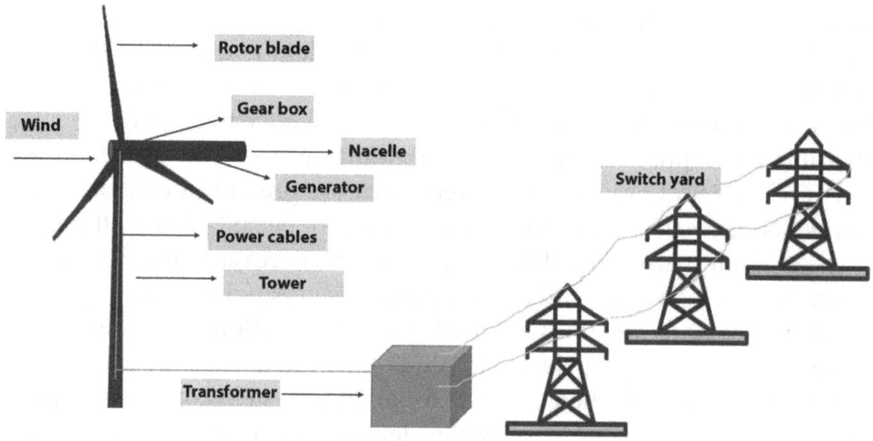

Figure 3.2 Diagrammatic representation of a wind turbine system.

According to NIWE, the nation has onshore capability of wind power of about 302 Global Wind, which is nearly eight times its current wind energy capacity. However, realizing this potential would require a significant investment in wind infrastructure and transmission facilities. The Indian government has been actively promoting wind energy through various policy initiatives and incentives [39].

In 2016, it launched the National Wind–Solar Hybrid Policy, which aims to promote the installation of wind–solar hybrid systems to optimize the utilization of land and grid infrastructure. In addition, the government has introduced various financial incentives such as generation-based incentives, accelerated depreciation, and concessional customs and excise duties to attract investments in wind power. The private sector has also been actively investing in wind power in India, given the favorable policy and regulatory environment [49]. Several domestic and international companies have set up wind power projects in the country, including Suzlon Energy, Tata Power, and Siemens Gamesa. In addition, several state governments have introduced their wind energy policies to attract investments and promote renewable energy. In spite of the significant possibility for wind power in India, several challenges need to be addressed to realize this potential. One of the biggest challenges is the intermittent nature of wind power, which forces it to be strenuous to combine into the grid [40]. The variability of wind energy could lead in accordance with grid instability and require backup power sources to be available. This, in turn, increases the affordability of wind energy and reduces the competitiveness compared with conventional sources of energy. Another challenge is the limited transmission infrastructure to transport wind power from remote locations to demand centers. This can result in congestion and curtailment of wind power, which reduces its utilization and profitability. Moreover, the lack of reliable wind data in certain regions of the country makes it difficult for the appraisal of the feasibility and potential of wind power projects correctly. Wind energy has enormous potential in India, but there are several limitations and challenges that need to be addressed to realize this potential [50]. Here are some of the significant limitations of wind energy in India.

3.4.1 Intermittency and Variability

Wind energy is inherently intermittent and variable, which makes it challenging to integrate into the grid. The output of wind power turbines varies with the speed of the wind, which can lead to fluctuations in the grid's frequency and voltage. This variability can lead to grid instability, require backup power sources, and increase the cost of wind power [51].

3.4.2 Land Acquisition

Wind power projects require significant land resources, particularly in the western and southern regions, where wind speeds are high and consistent. Land acquisition can be a challenging and time-consuming process, particularly in densely populated areas, where land is scarce and expensive [52].

3.4.3 Transmission Constraints

Wind power projects are often located in remote areas, far from the demand centers, which can create transmission constraints. The limited transmission infrastructure can result in congestion and curtailment of wind power, which reduces its utilization and profitability [53].

3.4.4 Limited Wind Resource Data

The availability of reliable wind resource data is mandatory for the assessment of feasibility and potential of wind energy projects accurately. However, there are several regions of the country where wind data is limited or not available, which makes it challenging to develop wind power projects in those areas [54].

3.4.5 Financing Constraints

Wind power projects require significant upfront investments, which can be a barrier for many developers, particularly smaller ones. Financing constraints, including high-interest rates and a lack of access to capital, can make it challenging to finance wind power projects [55].

3.4.6 Environmental and Social Impacts

Wind power projects can have significant environmental and social impacts, particularly on wildlife and local communities. The setting up of wind turbines can disrupt habitats of bats and birds and cause noise pollution. In addition, land acquisition can lead to the displacement of local communities and the loss of their livelihoods [56].

3.4.7 Policy and Regulatory Uncertainty

The government of India has introduced various policies as well as incentives for the promotion of wind power development. However, there is still policy and regulatory uncertainty, particularly around tariffs and grid connection charges, which can deter developers from investing in wind power [57]. Wind energy has enormous potential in India, but there are several limitations and challenges that need to be addressed to realize this potential fully. Addressing these limitations will require significant investments in wind infrastructure and transmission facilities as well as addressing the challenges of intermittency, land acquisition, transmission constraints, wind resource data, financing, environmental, and social impacts, and policy and regulatory uncertainty. Wind power can play a significant role in India's energy mix, with the regulatory framework, right policy, and investment, and assist the country in achieving its renewable energy targets [58].

3.5 Progress of Technology

Wind turbines are an essential component for the renewable sources of energy sector, providing a sustainable source of power without the environmental impact of fossil fuels. In the past few decades, the technology behind turbines has advanced significantly, with larger, more efficient turbines being developed that can generate greater amounts of power from the same amount of wind [41, 59].

3.5.1 Larger and More Efficient Turbines

One of the most significant advancements in wind turbine technology has been the development of more efficient and larger turbines. During the past days of wind energy, turbines were relatively small and produced only a few kilowatts of power. Today, however, turbines are much larger and more powerful, with some models capable of generating over 10 MW of electricity. The development of larger turbines has been driven by a number of factors. First, larger turbines are more efficient than smaller ones, as they can capture more wind energy and convert it into electricity. Additionally, larger turbines are better suited for use in offshore wind farms, where wind speeds are typically higher and there is more space available for installation [42, 60].

3.5.2 Advancements in Turbine Design

Along with the development of larger turbines, there have been significant advancements in turbine design over the past few decades. One of the most notable improvements has been the development of variable-pitch blades, which allow the turbine to adjust the angle of its blades based on the speed and direction of the wind. Variable pitch blades are particularly useful in areas where wind speeds can vary widely, as they allow the turbine to maintain a consistent level of power output regardless of the wind conditions (Figure 3.3). Additionally, variable pitch blades can help to extend the lifespan of the

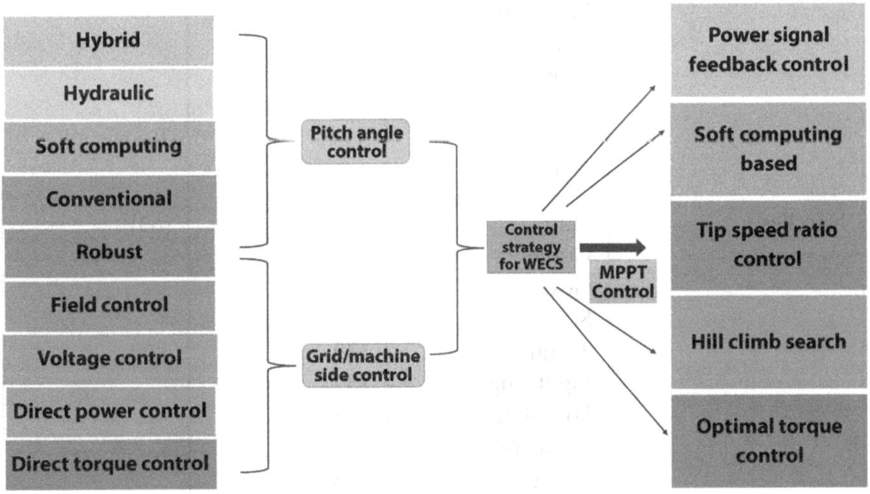

Figure 3.3 Schematic representation of different control strategies in wind energy control.

turbine, as they reduce the stress placed on the blades during high winds. Another important advancement in turbine design has been the development of direct-drive generators. Direct-drive generators eliminate the need for a gearbox, which can be a major source of maintenance and reliability issues in wind turbines. Direct-drive generators are also more efficient than traditional gearbox-based systems, as they eliminate the energy losses that occur during the transfer of power between the rotor and the generator [43, 61]. Table 3.1 represents the various frequencies of components used in wind technology.

Table 3.1 Keywords used in wind technology and their frequency range.

Domain	Keywords	Frequency	References
Generator or configuration	Stator	2,960	[62, 65]
	Tower	1,368	
	Grid	1,179	
	Battery	1,164	
	Gear	1,029	
	Fan	914	
	Hub	828	
	Pole	745	
	Cable	480	
	Sea	398	
Components or gearbox	Gear	1,923	[66]
	Oil	580	
	Gear box	574	
	Roller	565	
	Lubrication	483	
	Bolt	460	
	Flange	460	
	Cabin	459	
	Safety	333	
	Stator	261	
Nacelles	Hub	683	[63]
	Crane	206	
	Gear	197	
	Stator	163	
	Torque	139	
	Lightning	137	
	Foundation	133	
	Platform	129	
	Flange	128	
	Beam	66	

Table 3.2 Comparison of different kinds of pitch controls.

Controller	Reliability	Convergence speed	Complexity	Performance	Performance under rapid wind speed variation	Cost	References
Conventional controller	High	Slow	Low	Low	Low	Low	[64, 65]
Robust controller	Moderate high	High	High	Moderate high	High	High	[66]
Soft computing controller	High	Faster	Low	Moderate high	High	Moderate	[67]
Hybrid controller	High	Faster	Moderate	High	Very high	Moderate	[62, 63]

3.5.3 Improvements in Manufacturing and Installation

In addition to advancements in turbine design, there have been significant improvements in the manufacturing and installation of wind turbines over the past few decades. One of the most notable developments has been the use of modular construction techniques, which allow turbines to be assembled on site from pre-manufactured components. Modular construction techniques have several advantages over traditional manufacturing methods. First, they allow for greater quality control, as each component can be carefully inspected and tested before it is shipped to the installation site. Additionally, modular construction techniques can help to reduce the cost of manufacturing and installation, as they eliminate the need for expensive transportation and logistics. In addition, the main improvement in the manufacturing of turbines has been the use of advanced materials, such as carbon fiber and composites. These materials are stronger and lighter than traditional materials like steel, which allow for the development of larger and more efficient turbines. Table 3.2 shows different controller systems and their comparison with other components in terms of performance, speed, complexity etc.

3.6 Conclusion

Huge wind turbines are an essential part of the renewable energy landscape, addressing significant concerns and providing exciting potential in the future. With the ability to produce large amounts of electricity, these high towers can supply wind-derived renewable energy to cities and even small nations. Wind turbines help to lessen fossil fuel dependency as well as mitigate negative consequences for climate change by utilizing this limitless, free natural resource. There are significant concerns that need to be solved as wind turbine technology advances. One such problem is the requirement for effective and affordable storage technologies to control sporadic wind energy generation. The creation of cutting-edge energy storage technologies will provide a consistent and dependable supply of electricity even when the wind is not blowing at its best. The impacts which the large-scale wind farms would have on the environment also need to be carefully considered. To minimize any potential adverse effects on wildlife, migratory patterns, and aesthetics, proper siting and environmental studies are essential. Finding stability in the environmental preservation and production of renewable energy depends on ongoing study and innovation in these areas. A bright future is expected for the potential of large-scale wind turbines. Improvements in turbine production techniques, materials, and design offer the potential to lower prices and boost efficiency. Their performance can be improved through integration with smart grid technology and artificial intelligence, enabling optimized power generation and improved grid management. Offshore wind generation also offers a large unrealized promise. Even more energy can be produced while fewer problems over land use result from taking advantage of the steady and powerful winds at sea. Around the world, several locations are exploring and developing offshore wind farms.

References

1. Kirchhoff, T., Ramisch, K., Feucht, T., Reif, C., Suda, M., Visual evaluations of wind turbines: Judgments of scenic beauty or of moral desirability? *Landsc. Urban Plan.*, 226, 104509, 2022.
2. Watson, D., Guidelines for Landscape and Visual Impact Assessment (Second Edition). *URBAN Des. Int.*, 8, 97, 2003.
3. Torres Sibille, A., D., C., Visual Impact Assessment of Human Interventions on the Landscape: The Case of Wind Farms and Solar Power Plants. Doctoral Dissertation, Universitat Politècnica de València, 2011.
4. Jianu, O., Rosen, M.A., Naterer, G., Noise pollution prevention in wind turbines: Status and recent advances. *Sustainability*, 4, 1104, 2012.
5. Tonin, R., A Review of Wind Turbine-Generated Infrasound: Source, Measurement and Effect on Health. *Acoust. Aust.* 46, 69, 2018.
6. Chiu, C.H., Lung, S.C.C., Assessment of low-frequency noise from wind turbines under different weather conditions. *J. Environ. Heal. Sci. Eng.*, 18, 505, 2020.
7. Lukáčová, K., Badida, M., Moravec, M., Liptai, P., Particulate matters propagation by the surface winning. *Solid State Phenom.*, 244, 197, 2016.
8. Ma, B.B., Nystuen, J.A., Passive acoustic detection and measurement of rainfall at sea. *J. Atmos. Ocean. Technol.*, 22, 1225, 2005.
9. Kim, H.-S., An Analysis of the Sound Propagation between Rooms with Different Mediums. *J. Acoust. Soc. Korea.*, 32, 402, 2013.
10. Oud, M., Low-Frequency Noise: A Biophysical Phenomenon. In Congress Geluid Trillingen Luchtkwaliteit Gebied & Gebouw, 2012.
11. Chiu, C.H., Lung, S.C.C., Assessment of low-frequency noise from wind turbines under different weather conditions. *J. Environ. Heal. Sci. Eng.*, 18, 505, 2020.
12. Galparsoro, I., Menchaca, I., Garmendia, J.M., Borja, Á., Maldonado, A.D., Iglesias, G., Bald, J., Reviewing the ecological impacts of offshore wind farms. *npj Ocean Sustain.*, 1, 2022.
13. Joselin Herbert, G.M., Iniyan, S., Sreevalsan, E., Rajapandian, S., A review of wind energy technologies. *Renew. Sustain. Energy Rev.*, 11, 1117, 2007.

14. Ackermann, T., Söder, L., Wind energy technology and current status: a review. *Renew. Sustain. Energy Rev.*, 4, 315, 2000.
15. Staffell, I., Green, R., How does wind farm performance decline with age? *Renew. Energy*, 66, 775, 2014.
16. KUNZ, T.H., ARNETT, E.B., COOPER, B.M., ERICKSON, W.P., LARKIN, R.P., MABEE, T., MORRISON, M.L., STRICKLAND, M.D., SZEWCZAK, J.M., Assessing Impacts of Wind-Energy Development on Nocturnally Active Birds and Bats: A Guidance Document. *J. Wildl. Manage.*, 71, 2449, 2007.
17. Harrison-Atlas, D., Murphy, C., Schleifer, A., Grue, N., Temporal complementarity and value of wind-PV hybrid systems across the United States. *Renew. Energy*, 201, 111, 2022.
18. Albertani, R., Johnston, M.L., Clocker, K., Hu, C., Huso, M., Katzner, T., Maurer, W., Todorovic, S., Vang, J., A Heterogeneous System for Eagle Detection, Deterrent, and Wildlife Collision Detection for Wind Turbines. Office of Scientific and Technical Information (OSTI) 2021.
19. Sugihara, R.T., Pitt, W.C., Berentsen, A.R., Payne, C.G., Evaluation of the palatability and toxicity of candidate baits and toxicants for mongooses (Herpestes auropunctatus). *Eur. J. Wildl. Res.*, 64, 2018.
20. Bazilian, M., Outhred, H., Miller, A., Kimble, M., Opinion: An energy policy approach to climate change. *Energy Sustain. Dev.*, 14, 253, 2010.
21. Wagner, H. J., Mathur, J., *Introduction to Hydro Energy Systems: Basics, Technology and Operation*; Springer Science+Business Media, 2011.
22. Council, G. W. E., *Global Wind Energy Outlook,* 2008.
23. Chen, L., Ponta, F.L., Lago, L.I., Perspectives on innovative concepts in wind-power generation. *Energy Sustain. Dev.* 15, 398, 2011.
24. Solis, D., Book Review: Wind Energy – The Facts: a guide to the technology, economics and future of wind power by The European Wind Energy Association. *Int. J. Environ. Pollut.*, 43, 383, 2010.
25. Gawel, E., Purkus, A., Promoting the market and system integration of renewable energies through premium schemes - A case study of the German market premium. *Energy Policy*, 61, 599, 2013.
26. Ren, G., Liu, J., Wan, J., Guo, Y., Yu, D., Overview of wind power intermittency: Impacts, measurements, and mitigation solutions. *Appl. Energy*, 204, 47, 2017.
27. Martins, M.G., da Silva Arouche, T., Neto, A.F.G., da Cruz, J.N., da Costa, F.L.P., Fernandes, L.L., de Carvalho Junior, R.N., da Silva Costa, J.F., de Jesus Chaves Neto, A.M., Density functional theory for the thermodynamic gas-phase investigation of butanol biofuel and its isomers mixed with gasoline and ethanol. *J. Mol. Model.*, 27, 80, 2021.
28. Couture, T.D., Cory, K., Williams, E., A Policymaker's Guide to Feed-in Tariff Policy Design. Office of Scientific and Technical Information (OSTI) 2010.

29. Yang, Y., Wu, K., Long, H., Gao, J., Yan, X., Kato, T., Suzuoki, Y., Integrated electricity and heating demand-side management for wind power integration in China. *Energy*, 78, 235, 2014.
30. Rao, V., Bedford, A., Mokhtar, M., Howe, J. M., Power Profiling and Inherent Lag Prediction of a Wind Power Generating System for Its Integration to an Energy Storage System. International Conference on Systems, Man, and Cybernetics, *IEEE Publications*, 2013.
31. Negrete-Pincetic, M., Wang, G., Kowli, A., Pulgar-Painemal, H., Emerging Issues Due to the Integration of Wind Power in Competitive Electricity Markets. Power and Energy Conference At Illinois (PECI), *IEEE Publications*, 2010.
32. Dorsey-Palmateer, R., Effects of wind power intermittency on generation and emissions. *Electr. J.*, 32, 25, 2019.
33. Dehghani-Sanij, A.R., Al-Haq, A., Bastian, J., Luehr, G., Nathwani, J., Dusseault, M.B., Leonenko, Y., Assessment of current developments and future prospects of wind energy in Canada. *Sustain. Energy Technol. Assess.*, 50, 101819, 2022.
34. Proskuryakova, L.N., Ermolenko, G. V., The future of Russia's renewable energy sector: Trends, scenarios and policies. *Renew. Energy*, 143, 1670, 2019.
35. Nagtode, V.S., Cardoza, C., Yasin, H.K.A., Mali, S.N., Tambe, S.M., Roy, P., Singh, K., Goel, A., Amin, P.D., Thorat, B.R., Cruz, J.N., Pratap, A.P., Green Surfactants (Biosurfactants): A Petroleum-Free Substitute for Sustainability-Comparison, Applications, Market, and Future Prospects. *ACS Omega.*, 8, 11674, 2023.
36. Sadorsky, P., Wind energy for sustainable development: Driving factors and future outlook. *J. Clean. Prod.* 289, 125779, 2021.
37. Sahu, B.K., Wind energy developments and policies in China: A short review. *Renew. Sustain. Energy Rev.*, 81, 1393, 2018.
38. Meibom, P., Weber, C., Barth, R., Brand, H., Operational costs induced by fluctuating wind power production in Germany and Scandinavia. *IET Renew. Power Gener.*, 3, 75, 2009.
39. Chang, W.-Y., A Literature Review of Wind Forecasting Methods. *J. Power Energy Eng.*, 02, 161, 2014.
40. Zhang, P., Jia, N., Le, C., Ding, H., Penetrating-levelling tests on bucket foundation with inner compartments for offshore wind turbines in sand. *Mar. Struct.*, 82, 103143, 2022.
41. Lee, Y.S., González, J.A., Lee, J.H., Kim, Y. Il, Park, K.C., Han, S., Structural topology optimization of the transition piece for an offshore wind turbine with jacket foundation. *Renew. Energy*, 85, 1214, 2016.
42. Roddier, D., Cermelli, C., Aubault, A., Weinstein, A., WindFloat: A floating foundation for offshore wind turbines. *J. Renew. Sustain. Energy*, 2, 2010.
43. LIANG, F., ZHENG, H., ZHANG, H., On the pile tension capacity of scoured tripod foundation supporting offshore wind turbines. *Appl. Ocean Res.*, 102, 102323, 2020.

44. Satpute, A.V., Kumar, E.V., Current scenario of wind power in India, government policies, initiatives, status and challenges. *Int. J. Energy Sect. Manag.*, 15, 209, 2021.
45. Kong, K., Dyer, K., Payne, C., Hamerton, I., Weaver, P.M., Progress and Trends in Damage Detection Methods, Maintenance, and Data-driven Monitoring of Wind Turbine Blades – A Review. *Renew. Energy Focus*, 44, 390, 2023.
46. Wang, M., Wang, C., Hnydiuk-Stefan, A., Feng, S., Atilla, I., Li, Z., Recent progress on reliability analysis of offshore wind turbine support structures considering digital twin solutions. *Ocean Eng.*, 232, 109168, 2021.
47. Dorian, J.P., Franssen, H.T., Simbeck, D.R., Global challenges in energy. *Energy Policy*, 34, 1984, 2006.
48. Ohunakin, O.S., Akinnawonu, O.O., Assessment of wind energy potential and the economics of wind power generation in Jos, Plateau State, Nigeria. *Energy Sustain. Dev.*, 16, 78, 2012.
49. Come Zebra, E.I., van der Windt, H.J., Nhumaio, G., Faaij, A.P.C., A review of hybrid renewable energy systems in mini-grids for off-grid electrification in developing countries. *Renew. Sustain. Energy Rev.*, 144, 111036, 2021.
50. Ghaffour, N., Bundschuh, J., Mahmoudi, H., Goosen, M.F.A., Renewable energy-driven desalination technologies: A comprehensive review on challenges and potential applications of integrated systems. *Desalination*, 356, 94, 2015.
51. Gao, Y., Ma, S., Wang, T., Wang, T., Gong, Y., Peng, F., Tsunekawa, A., Assessing the wind energy potential of China in considering its variability/intermittency. *Energy Convers. Manag.* 226, 113580, 2020.
52. Kumar, A., Pal, D., Kar, S.K., Mishra, S.K., Bansal, R., An overview of wind energy development and policy initiatives in India. *Clean Technol. Environ. Policy*, 24, 1337, 2022.
53. Hozouri, M.A., Abbaspour, A., Fotuhi-Firuzabad, M., Moeini-Aghtaie, M., On the use of pumped storage for wind energy maximization in transmission-constrained power systems. *IEEE Trans. Power Syst.*, 30, 1017, 2015.
54. Jung, C., Schindler, D., Laible, J., National and global wind resource assessment under six wind turbine installation scenarios. *Energy Convers. Manag.*, 156, 403, 2018.
55. Kerr, W.R., Nanda, R., Financing Constraints and Entrepreneurship. *SSRN Electron. J.*, 2011.
56. Jaber, S., Environmental Impacts of Wind Energy. *J. Clean Energy Technol.*, 251, 2014.
57. Fabrizio, K.R., The effect of regulatory uncertainty on investment: Evidence from renewable energy generation. *J. Law, Econ. Organ.*, 29, 765, 2013.
58. Ghaffour, N., Bundschuh, J., Mahmoudi, H., Goosen, M.F.A., Renewable energy-driven desalination technologies: A comprehensive review on challenges and potential applications of integrated systems. *Desalination*, 356, 94, 2015.

59. Jonkman, J., Butterfield, S., Musial, W., Scott, G., Definition of a 5-MW reference wind turbine for offshore system development. Office of Scientific and Technical Information (OSTI) 2009.
60. Eriksson, S., Bernhoff, H., Leijon, M., Evaluation of different turbine concepts for wind power. *Renew. Sustain. Energy Rev.*, 12, 1419, 2008.
61. Asim, T., Islam, S.Z., Hemmati, A., Khalid, M.S.U., A Review of Recent Advancements in Offshore Wind Turbine Technology. *Energies*, 15, 579, 2022.
62. Elgendi, M., AlMallahi, M., Abdelkhalig, A., Selim, M.Y.E., A review of wind turbines in complex terrain. *Int. J. Thermofluids.*, 17, 100289, 2023.
63. Khalid, M.Y., Arif, Z.U., Hossain, M., Umer, R., Recycling of wind turbine blades through modern recycling technologies: A road to zero waste. *Renew. Energy Focus.*, 44, 373, 2023.
64. Wilberforce, T., Olabi, A.G., Sayed, E.T., Alalmi, A.H., Abdelkareem, M.A., Wind turbine concepts for domestic wind power generation at low wind quality sites. *J. Clean. Prod.*, 394, 136137, 2023.
65. Pandit, R., Astolfi, D., Hong, J., Infield, D., Santos, M., SCADA data for wind turbine data-driven condition/performance monitoring: A review on state-of-art, challenges and future trends. *Wind Eng.*, 47, 422, 2023.
66. Timmer, W.A., Bak, C., Aerodynamic characteristics of wind turbine blade airfoils. In: *Advances in Wind Turbine Blade Design and Materials*. pp. 129–167. Elsevier, 2023.
67. Yahya, W., Ziming, K., Juan, W., Al-Nehari, M., Tengyu, L., Qichao, R., Alhayani, B., Study the influence of using guide vanes blades on the performance of cross-flow wind turbine. *Appl. Nanosci.*, 13, 1115, 2023.
68. Xiong, J., Peng, T., Tao, Z., Zhang, C., Song, S., Nazir, M.S., A dual-scale deep learning model based on ELM-BiLSTM and improved reptile search algorithm for wind power prediction. *Energy*, 266, 126419, 2023.

4

Wind Hybrid Power Technologies

Momina, Haq Nawaz Bhatti* and Amina Khan

Department of Chemistry, University of Agriculture, Faisalabad, Pakistan

Abstract

The global usage of renewables is expanding rapidly these days. Wind power accounts for the majority of the production of clean energy worldwide. Due to the turbulence in wind speed, there may be fluctuations in wind power generation. To compensate this loss, windmill is integrated with different energy sources like solar, diesel generator, hydrogen system, hydroelectric, and PV power plants. These hybrid energy systems can work as an isolated system or with a grid system. The advantages of the abovementioned electrical networks include reliable supply, reduced carbon emissions, enhanced power quality, reliability, and work opportunities for local people.

Keywords: Renewable energy, hybrid system, power generation

4.1 Introduction

The need for higher living standards in emerging nations has grown over time as society and the economy have evolved. This has increased the awareness of environmental contamination. Renewable energy, which is regarded as a clean energy, can be a substitute of fossil fuels that pollute the environment and are not recyclable in order to provide enough power to support people's daily needs and societal progress [1]. The global usage of renewable energy is expanding rapidly these days. According to the 2018 Renewable Capacity Statistics, the global renewable energy-generating capacity reached 2,179 GW at the end of 2017. With 1,152 GW, hydro had the highest proportion. Wind, solar, biomass, geothermal, and tidal power are among the energy sources. Renewable sources of electricity (wind,

**Corresponding author*: hnbhatti2005@yahoo.com

Inamuddin, Tariq Altalhi and Mohammad Luqman (eds.) Wind Energy Storage and Conversion: From Basics to Utilities, (63–82) © 2024 Scrivener Publishing LLC

sun, and water) are reasonably abundant in certain distant places, but the national grid has not been linked to these areas [2]. Wind power has grown quickly in the past decade to become the most important renewable energy source [3].

Wind power accounts for the majority of the production of clean energy worldwide. The wind revealed the vast potential of diverse places in terms of wind energy because mountain chains along coastlines provide a natural gateway that improves wind stability. Most coastal locations benefit from being close to areas where power is in high demand [4].

Hybrid power plants integrate two or more types of energy production, storage, or end-use technologies, so they provide significant advantages over single-source systems. Versatility in our daily lives may be called the spice of life; hence, why restrict ourselves to just a single power source or storage option? In these circumstances, hybrid power plants are an appropriate answer since they may provide significance increase in efficiency and cost savings while also being adaptable to varied end-user needs [5]. Hybrid energy systems have been built to generate power from many sources. In this chapter, we will discuss about the wind hybrid power systems in which wind turbines are associated with solar, hydro, diesel, PV, or hydrogen systems [6].

As clean energy, modern technology progresses and petroleum costs rise, hybrid renewable energy systems are growing increasingly viable as standalone power systems for delivering electricity in remote places. A hybrid power system, commonly referred to as hybrid power, integrates two or more forms of renewable energy in order to improve the system's effectiveness and supply balance [7]. The advantages of the abovementioned electrical networks, including reliable supply, reduced carbon emissions, enhanced power quality, reliability, and work opportunities for local people, are the primary motivations for their implementation. Due to the intrinsically intermittent nature of green energy sources, hybrid combinations involving two or more renewable energy sources are utilized [8].

4.2 Types of Hybrid Power Systems

Following are the types of hybrid power systems:

a) Isolated hybrid grid system
   If a hybrid system possesses three sub-units—power production unit, power distribution unit and power demand—then,

it is called an isolated hybrid micro-grid system. The most crucial aspect of a micro-grid is maintaining a constant supply of electricity based on consumer demand, which is difficult to anticipate, and energy output may vary depending on the availability of renewable energy sources, i.e., wind. The key issues with micro-grids include fluctuations in voltage, harmonic current, and voltage instability [9].

b) Grid-tied hybrid micro-grid system
   When a hybrid micro-grid is linked to the main grid, the system is referred as a grid-tied hybrid micro-grid system [3]. Grid-tied mode refers to the connection of the micro-grid to the grid utility. This system copes up when power fluctuation arises due to the varying availability of wind and fulfill the demand of users by providing electricity. When a micro-grid is separated from the grid utility, it enters the standalone mode. The micro-grid immediately shifts to autonomous mode during fault diagnosis [10].

4.3 Wind Hybrid Power Technologies

Wind farms are integrated with different power-generating systems. The most commonly used wind hybrid power technologies, their working components, and their advantages and disadvantages are given below.

4.3.1 Wind Diesel Hybrid Power Technology

Diesel generators are primarily used as backup power supplies in telecommunications, public buildings, hospitals, and other civic structures, as impartial armed and marine voltage plants, and as a trustworthy isolated source of electricity for landmasses or distant towns located far away from the main electric grid [13]. Because synchronous generators are used in traditional generators powered by internal combustion engines, static speed action is essential. This system has a poor efficacy while under short voltage. It is not vital in emergency-run setups, but it is critical in constantly operated units where the consumption of fuel is a key economic and logistical factor. Furthermore, running a diesel generator at synchronized speed limits the engine's lifespan due to partial combustion of hydrocarbons, and thus a supplementary dumping power is necessary to optimize the ignition procedure. By using the load-adaptive, changeable-speed setup of the generator, we can increase efficacy and combustibility during low load situations [14]. Hybrid wind–diesel island power stations have been deployed. The use of wind energy reduces fuel consumption, while the diesel generator ensures the stability of voltage supply to the consumer. The diesel generator cannot be totally excluded since wind turbines are unreliable power sources that require massive energy storage to ensure electricity in low wind circumstances [7]. The classification of wind diesel hybrid system has been discussed in Table 4.1.

Figure 4.1 depicts the components of a wind diesel hybrid system.

Table 4.1 Wind diesel classification.

Penetration class	Operating characteristics	Instantaneous penetration
Low	Diesel runs full time. Wind power reduces net load on diesel. All wind energy goes to primary load. No supervisory control system	<50%
Medium	Diesel runs full time. At high wind power, second wind load is dispatched to ensure sufficient diesel loading or wind generation is curtailed. Requires a relatively simple control system	50%–100%
High	Diesel may be shut down during high wind availability. Auxiliary components required to regulate voltage and frequency. Require sophisticated control system	100%–400%

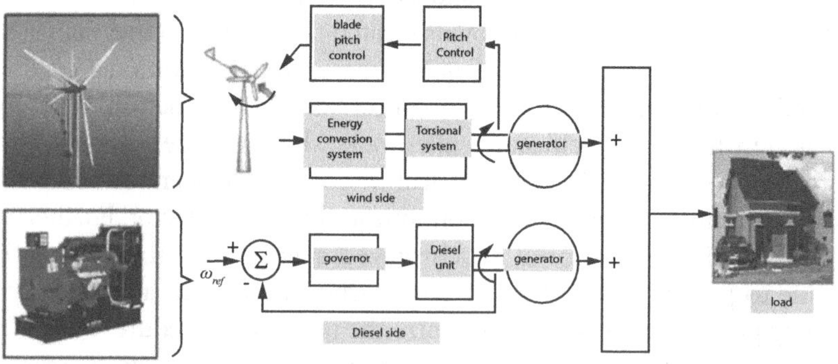

Figure 4.1 Schematic view of the wind–diesel power technology.

- Wind turbines
- Diesel generators
- Additional dump load
- Storage batteries
- Power control system [11]

Wind turbines: The majority of tiny windmills manufactured nowadays are parallel-axis, windy turbines with three or more blades. They are often made of a composite substance (fiberglass). The frame of the turbine refers to the structure upon which the tail and rotor engine are fastened. The amount of energy generated is determined by the diameter of the blade. The circumference of the rotor determines the swept area. The tail helps the windmill point in the right direction [12]. Tiny wind mills are installed on the upper edge of the tower since the air speed rises as the tower gets higher, as does the power generated by the turbines. As an example, when we increase the total length of the wind mill tower from 60 to 100 ft, it raises the amount of power produced by 25% and increases the price by 10% of the overall price of the entire system [13].

Diesel generator: The main function of a diesel generator is to provide backup power. If the batteries reach their lowest permissible level or the home wants extra power, the diesel engine drives in. It can also be run manually. The engine adjusts to the changes in the home load over the day. The power plant also charges the storage devices by producing more electricity than what the home consumes on a consistent basis. Whenever the diesel engine is turned on because of shortage of battery voltage, it indicates an extensive energy shortage in the entire system, which may not be properly rectified by a small period of output. The power control system makes no deduction of how long the generator has been running while making its judgment. As a result, there is a need for an automated controller to utilize the generator more efficiently to handle fluctuations in energy supply from its turbine and battery backup [14].

Additional dump load: This unit is mainly used to guarantee that any surplus energy delivered to the system could be safely dispersed. The dump load is utilized to raise the temperature of the home hot-water cylinder, which saves power but restricts the use of alternate heat sources such as a solar heater. Nevertheless, it can be far preferable when the heat energy can be stored and kept as a contributor for the house's current needs, reducing the total power that has to be provided by the diesel generators. Another essential role of the dump load is to guarantee that the wind turbine is always loaded. When there is no method to distribute extra power, the turbine's filling may fall, lowering resistive torque and leading the blades to over-speed [14].

Storage batteries: The batteries considerably contribute to the entire initial expenditure of the system, which is about similar to the diesel engine, and maintenance is typically considerable. As a result, the actual capacity of the storage devices is a critical economic and operational concern [15]. A distinctive regular schedule finds the system dragging power out of the storage batteries in the morning hours of the day, shortly before a midday meal and late at night because of the regular sequences of wind speed and home voltage consumption. Battery packs are then recharged in the evening, when the wind is at its strongest.

Power control system: The main function of a power control system is to distribute power between different components carefully. It also ensures that there is a stable power supply towards the load and grid.

When considering the quantity of wind which is incorporated in the grid system, a method of categorization is employed. A system is termed a highly penetration structure if the quantity of wind supplied at one given moment exceeds 100% of the overall quantity of energy generated. Low penetration systems have lower than 50% peak instant saturation, whereas medium diffusion schemes have 50% to 100% of the power generated from the windmill at one particular moment. Low and medium penetration systems are well-established systems. Nevertheless, high penetration systems continue to have several issues, particularly when fixed, with its ability to run when the diesel mode is off.

Before delving into the different wind–diesel operating modes, it is a good idea to define the sorts of situations to which they must adapt. The response of wind turbines to realistic, unpredictable wind is the key factor of a wind–diesel design. Rapid changes in wind speed, lasting from a few seconds to almost 10 min, cause significant fluctuations in engine power, which can substantially disrupt system performance. In addition, during strong wind times, extra electricity may need to be discharged, and temporary storing may be required. Fuel efficiency dictates that the diesel generator must be switched off if the mean energy produced by the windmill, compared to the demand, is more. Diesel fuel consumption reduces dramatically under partial load, and engine damage can be severe [16]. The basic premise of the wind diesel dilemma is the tension among the need to turn off the engine, avoiding repeated starts and low partial load functioning [16].

Following are the types of operating conditions of WDHPT:

- Diesel-only mode (DO): When operating in DO mode, the windmill's maximum power is scarcely more than the entire load required by the consumer. At this instance, the diesel power plants (DPP) never shut down and continue to provide the active and reactive electricity required by the customer's demand. However, DG fuel consumption can be lowered depending on the quantity of voltage generated by the windmill (usually disconnected) and specific aspects of the diesel generator.

- Wind–diesel mode (WD): In WD mode of operation, the windmill output is typically equivalent to the customer's demand, and wind generators deliver active power in conjunction with DG(s). Wind gusts fluctuate from a moment to the next due to turbulence in the wind; thus, instantaneous wind turbine generator (WTG) power fluctuations usually go beyond the customers' demand. In these circumstances, the following operating ways are feasible:
  (a) A diesel engine can operate constantly.
  (b) The diesel engine can be shut down and resumed based on immediate wind energy and according to load needs. Constantly using the diesel engine reduces the load needed, increasing the previously discussed diesel operating expenses.
- Wind-only mode (WO): WO mode uses no fuel because the WTG power is frequently much higher than what is required for load, and the diesel-powered generators are not operating. In this situation, utilizing batteries for energy storage might enhance the entire system's economics [16]. The advantages and disadvantages of wind diesel hybrid power system are given in Table 4.2.

Table 4.2 Advantages and disadvantages of wind–diesel hybrid power technology.

Advantages	Disadvantages
• Reliable power-generating system since a generator provides the backup power • Reduce fuel usage because wind provides electricity most of the time • Less emission of greenhouse gases • Low operating cost because wind power is free and fuel consumption is reduced	• High capital cost than a single diesel generator • Wind turbines require frequent maintenance • The speed of wind may vary which can disrupt the power supply • Cannot be suitable for large-scale power generation systems • High fuel costs

4.3.2 Wind Solar Hybrid Power Technology (WSHPT)

A hybrid system that combines a solar array and a windmill is a form of hybrid power system. This setup will provide the most power if the speed of the wind is high, and solar panels will generate the most power in the peak summer season. In a WSHPT, the wind turbine produces energy whenever the appropriate wind speed is accessible, and the solar array produces electricity when the required quality of solar power is present. This sort of hybrid power plant requires less upkeep but is more dependable and has a longer lifespan [17].

Figure 4.2 shows the components of a wind solar hybrid system.

Thermal–solar array: A solar array captures sunlight. This sunlight is used to boil water and convert it into steam. The steam is then used to produce electricity [18].

Wind turbine: It is a device that efficiently creates electrical power by using wind power. The pressure, amount, and the speed of the wind can all be utilized to create electrical energy (AC) [10].

Batteries: The electrical power created by the hybrid power technology must be stored so that batteries are necessary. These batteries store energy

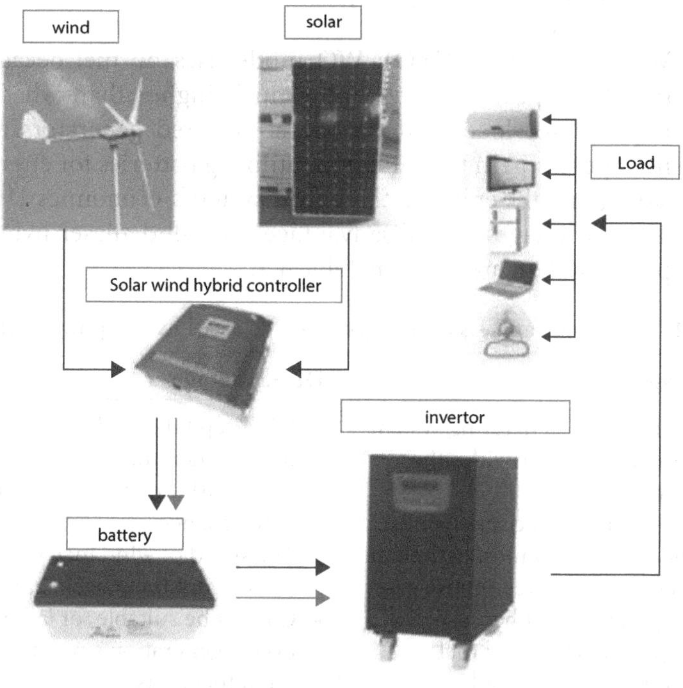

Figure 4.2 Schematic view of the wind–solar power technology.

and offer backup power during periods of insufficient sunlight or wind energy [16]. As a result, batteries must meet the following requirements:

- must withstand multiple charging and discharging cycles
- must have less self-destruction
- can function effectively within the parameters stated

The level of use and ambient temperature of the battery have an impact on its ability to store power. Its size is based on how much energy it uses and is rated in Ampere-hours (Ah). The following equation may be used to compute the battery's storage capacity if the mean amount of electricity used per hour is known and the average value for the life of the battery is also known:

$$BC = 2 \times f \times \frac{W}{Vbatt} \tag{4.1}$$

In the above equation,
BC = battery capacity
f = factor for reverse
W = daily energy
Vbatt = DC voltage of the system

Furthermore, the rating of the battery in Ah is determined as follows:

$$\text{rating of battery} = \frac{\text{daily energy consumption in KW}}{\text{battery rating in Ah at a specified voltage}} \tag{4.2}$$

Inverter: Though the majority of the power used is alternating current (AC), batteries store and supply direct current (DC). A device called an inverter transforms direct current into alternating current. Therefore, it is placed between the storage systems and the AC load. It might be electronic or analogue. Utilizing MOSFET (metal–oxide–semiconductor field-effect transistor Technology), the electronic inverter, which makes use of a microprocessor, offers improved efficiency. As a result, the system's construction expenditure rises. In order to construct the framework, an analogue inverter is investigated. The major source of energy for a home, AC (120 V, 1,500 W, 50 Hz), is converted from DC energy with the aid of an inverter. The electrical output portion can be utilized for coupling AC

Table 4.3 Advantages and disadvantages of wind–solar hybrid power technology.

Advantages	Disadvantages
• Wind energy help to reduce the cost of generating electricity • Clean source of energy since no emission of greenhouse gas • Both sources complement each other • Can be planted for power generation for large areas	• Both solar and wind are variable sources of energy. • Highly complex infrastructure than a standalone renewable energy system. • A large area is required for their plantation. • Batteries are required for the storage of surplus power.

loads [19]. Furthermore, the inverters include a few built-in safety features, namely:

- Restarting automatically: Automatic switch-off is possible if the power supply or array output is sufficiently high or sufficiently low.
- To handle an overloaded or short circuit condition, a protecting strategy is offered.
- For high-quality output, inverters with the specified properties are employed [16].
- There is protection against low battery, the drip and continual charging method.
- The load status indicator is present.
- Overload precautions are available [20].

Maximum charged battery selection: This is the circuit that has been created. The goal of maximum charged battery selection is to identify the maximum amount of electrical power that can be stored in each battery. Maximum charged batter selection is frequently connected within the inverter's circuit. As a result, the maximum charged battery selection is surrounded by the inverter block. This circuit may be built using the transistor series in conjunction with resistors, capacitors, and diodes [21]. The advantages and disadvantages of this system are given in the Table 4.3.

4.3.3 Wind Hydrogen Hybrid Power Technology (WHHPT)

Wind and hydrogen fuel are regarded as clean energy resources that are gaining popularity around the world. In accordance with an EU estimate, hydrogen is expected to be responsible for 24% of the total energy supply in 2050. The Chinese administration has also prioritized the development of the hydrogen-based economy. Among the most appealing possibilities for sustainable applications is the generation of hydrogen in conjunction with fuel cells for use [22].

Figure 4.3 describes the components of a wind hydrogen hybrid power system.

Wind turbines: Wind turbines generate electric power from the kinetic energy of wind. The amount of electricity produced depends upon the size of the turbine and availability of wind sources.

Electrolyzer: The electric energy is then transferred to an electrolyzer which electrolyzes the water into hydrogen and oxygen. Oxygen is released into the environment, and the excess of electricity is stored in hydrogen as electrochemical energy [23]. The minimal capacity of the electrolyzer in a WHHS should be cautiously determined because it has varying operational power constraints. If a lower nominal capacity electrolyzer is used, it will operate at relatively slow winds and create hydrogen progressively owing to its lower minimal power usage limit. Nevertheless, the maximum operational power restriction is likewise modest in this situation. Because the electrolyzer cannot operate over its greatest operational power constraint at high winds, the majority of the wind power will be squandered. If, on the contrary hand, a greater nominal power electrolyzer is used, the electrolyzer will only function at extreme winds, which are typically uncommon [24].

Storage tanks: Storage tanks are used to store hydrogen. Hydrogen is present in a highly compressed state so tanks must be able to accommodate

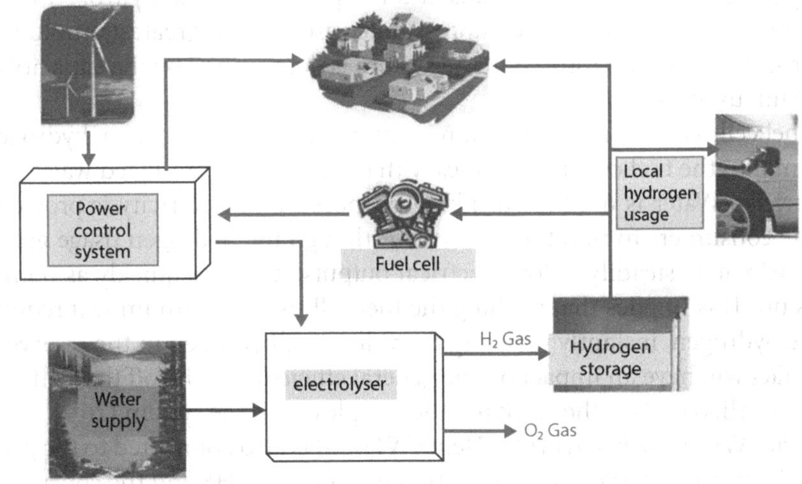

Figure 4.3 Schematic view of the wind-hydrogen hybrid technology.

that pressure to avoid any disaster and store hydrogen for future use [23]. The hydrogen created by the electrolyzer is stored in the storage tank and utilized in the fuel cell stack as demanded. We suppose that the level of compression in the reserve container can reach 30 bar (3,000 kPa). Hydrogen may be considered to be an ideal gas at a critical pressure (1.3 MPa), and general gas rules can be applied. Beyond this threshold pressure (>1.3 MPa), hydrogen behaves like a true gas. There is a difference between the true volume of gas and the estimated volume utilizing ideal gas laws at elevated pressures for hydrogen. The coefficient of compressibility (Z) is commonly used to explain this variation. The Boyle–Charles equation for an ideal gas is stated as follows:

$$PV = nRT \tag{4.3}$$

where P is the pressure, V is the volume, n is the number of moles, R is the general gas constant, and T is temperature. Equation 4.3 at high pressure is stated as follows:

$$PV = ZnRT \tag{4.4}$$

To compute the quantity of hydrogen gas in the storage container, we utilize Equation 4.4. When the container's pressure is less than the critical pressure of hydrogen gas, i.e., 1.3 MPa, we use the compressibility factor $Z = 1$ and Equation 4.4 becomes equal with Equation 4.3. However, when the pressure within the container surpasses 1.3 MPa, the ambient temperature of the hydrogen falls and the pressure of the hydrogen rises; the impact of the factor of compressibility grows. The greatest amount of hydrogen that may be stored in the system is determined by the capacity of the tank used for storage [25].

Fuel cell: When there is a requirement of electricity in an area, hydrogen is sent into the fuel cell. It combines with oxygen gas to produced water and electricity. Water is used for drinking purposes, and electricity is provided to the consumers to fulfill their needs. Although the hydrogen usage grows approximately steadily at low electrical outputs, it climbs quickly as output goes up. This implies that pushing the fuel cell to its maximum will require more hydrogen, instantly reducing the volume of hydrogen in the reservoir. This fact will have an impact on the general efficiency of the off the grid system as well, based on the stack number employed in the system [26].

Grid: Wind–hydro hybrid system (WHHS) is also connected to the grid. The electric grid serves as a connection between WHHS and the consumer. It provides electricity to the consumer or store if available in excess [27].

Table 4.4 Advantages and disadvantages of wind–hydrogen hybrid power technology.

Advantages	Disadvantages
• Highly effective because of their high density and safety • Pollution-free process • Less generation costs	• High investment of system configuration • Expensive storage tanks • Loss of energy during conversion and storage process

Power control systems: The power control system analyzes how all the components of WHHPS are working. It assures that there is a stable power flow between these components [28]. The advantages and disadvantages of this system are given in the Table 4.4.

4.3.4 Wind–Hydro Hybrid Power Technology (WHHPT)

A WHHPT uses a mix of wind farms and hydroelectric turbines to generate electricity; this sort of pairing is a long-lasting experiment and exploration facility. The world's first wind–hydro hybrid power station was built on the island of El Hierro [29]. Wind energy varies in nature, and when the wind rate is particularly strong, there is a danger of damage to wind blades. However, the production of wind turbines is undesirable when there is little wind. To address this issue, an integrated power source is employed to generate electricity. If wind power is unavailable in an integrated wind hydropower plant, hydropower is employed to solve the scheduling challenge. However, the effectiveness of a small hydroelectric dam is determined by the quantity of water available. Wind–hydro hybrid power technology is an array that is commonly utilized on islands with interconnecting systems, and wind energy is the primary energy source. It is designed to increase the use of renewable electricity as well as lower the (levelized cost of energy) LCOE, overall power shortfall, and energy generated by traditional hydroelectric dams [30].

Figure 4.4 shows the components of a wind hydro hybrid power system.

Wind turbine: Wind turbines generate electric power from the kinetic energy of wind. The amount of electricity produced depends upon the size of the turbine and availability of wind sources. The wind turbine might be manually linked to the pump through a gearbox or electrically by converting wind power to electric energy. Both forms have unique properties; nevertheless, the electrical variant is better recognized and employed.

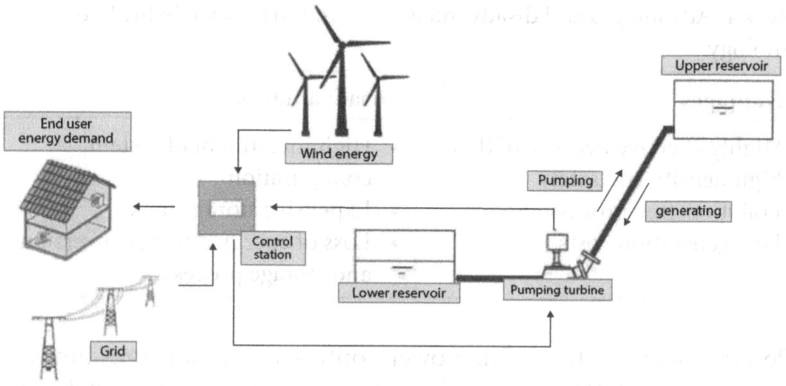

Figure 4.4 Schematic diagram of wind-hydro hybrid technology.

This is owing to the substantial loss of energy and mechanical variations that might happen [31].

Hydro turbines: These turbines generate electricity by using the potential energy of water stored in the upper reservoir. When water moves down from reservoir, its potential energy changes into kinetic energy and this kinetic energy is used to turn turbines. By the rotation of turbines, electricity is produced [32].

Water reservoirs: Two water reservoirs are used in WHHPS. One is called upper reservoir, while the other is lower reservoir. It is preferable for the hydrothermal (HT) to function 24 h a day, and the upper reservoir volume needs to be constructed in such a manner that the HT can work throughout the day as well as the night. It is preferable for the initial quantity of water reserve in the upper reservoir to be as high as feasible. The optimum performance of the system is not only taken into account at the phase of design and component dimensions, but it also involves optimum operations and time planning [33].

Pumping system: This system is used to pump water from the lower reservoir to the upper reservoir. Once the water falls down from the upper reservoir to generate electricity, it needs to be driven back toward the upper reservoir for storage [2].

Energy storage system: When the wind speed is high and the generated electric power is greater than what the consumers need, then an extra amount of electricity is used to push the water uphill from the lower reservoir to the upper reservoir. This helps to compensate the decrease in electric power due to fluctuation and instability in wind speed [34].

Regulators: Three types of regulators may be utilized in the WHHPS: storm controllers, (high voltage DC) HVDC controllers, and load following

Table 4.5 Advantages and disadvantages of wind–hydro hybrid power technology.

Advantages	Disadvantages
• Can be used to smooth offshore wind energy changes • Regulate electricity demand • Reduce the cost of electricity • Reduce wind power risks	• High capital cost • Output is weather dependent • High maintenance costs (water and wind turbines require continual maintenance) • This hybrid system is location specific (area where the wind speed is high and water is available is usually selected).

controllers. These regulators help to maintain the energy supply to the consumers. It also monitors how all parts of WHHPS are working [35].

To profit from a wind–hydro hybrid system and assist farms with irrigation, efficient management of electricity may be conducted in agriculture micro-grids. Increased wind penetration is an additional approach for making the system economically feasible. This will increase PHES's earnings and, consequently, its payback period [36]. Keeping this in mind, it is vital to constantly review if the system is operating at its best since each function must be tailored to a given amount of power. When comparing the WHHPS system to traditional reservoirs, it is concluded that, while WHHPS has a greater initial cost, it has a lower operational cost, environmental effect, and flooded area. The system's initial expense is heavily influenced by wind energy availability and plant building area [37]. The advantages and disadvantages of this system are given in the Table 4.5.

4.3.5 Wind-Photovoltaic (PV) Hybrid Power Technology

Because of the weather conditions, photovoltaic (PV) systems are unable to produce electricity throughout the day [38]. A hybrid renewable generation system (HRGS) idea has been proposed in order to preserve energy production consistency. Although wind farms and PV plants are significant sources of electricity, their energy output is variable and highly reliant on weather conditions, causing a mismatch in renewable electrical supply and customer demand for electricity [39].

Figure 4.5 shows the components of PV wind technology.

Wind farm: A wind turbine is a mechanical device that efficiently creates electricity by using wind power. The amount, pressure, and speed of the wind can all be utilized to create electrical energy [6].

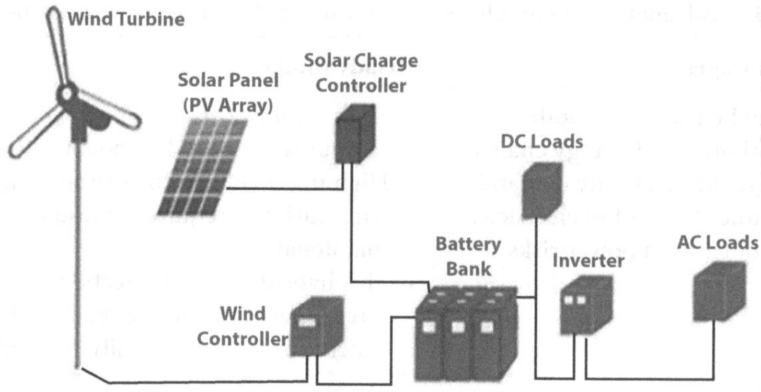

Figure 4.5 Schematic diagram of wind–PV hybrid technology.

PV plant: A PV plant consists of photovoltaic cells attached with a solar panel. These photovoltaic cells are composed of semi-conductor materials such as silicon. Unlike thermal solar panels, PV cell converts sunlight into electricity [40].

Energy storage unit: If the electrical outputs of the windmill and PV setup surpass the capacity of the power transmission channels, the residual electricity is transformed into thermal power by energy harvesting (EH) and retained in the thermal energy storage (TES) devices. In PV, EH and TES can be used for this purpose. The melting salt in the cold container gets pumped towards the EH and heated by surplus energy before being stored in the hot tank. The TES system will discharge heat energy into the power block (PB) to create electricity to meet the power shortfall. The melting salt in the heated tank is poured into the heat exchangers to transform into thermal energy and turn water into steam, and then the steam is pushed into the PB to push steam turbines and create electricity. In addition to reducing wind curtailment and solar curtailment and allowing for smoother wind and PV production fluctuations, the combined functioning of the EH and TES systems also allows for the adjustment of supply periods to maximize the capacity of the transmission channels [20].

Power control system: The power control system analyzes how all the components of WHHPS are working. It assures that there is stable power flow between these components [41].

Converter: The power produced by the windmill must be rectified and exchanged a DC (direct current) bus with the photovoltaic plant. The DC power is transformed to AC (alternating current) by the power inverter and amplified by the transformer before being incorporated into the power grid [42].

Table 4.6 Advantages and disadvantages of wind–PV hybrid power technology.

Advantages	Disadvantages
• Highly efficient power-generating system • Convert sunlight directly into electric energy • Lower the electricity bill	• Since PV cells consists of semiconductor materials, their capital costs is high. • Less power storage ability. • Large area is required for their plantation. • High maintenance costs.

The uncertain and unpredictable nature of PV and wind turbines can lead to a variety of turbulence issues, including variability in voltage and deviations of frequency. To solve this problem, the dimensions of PV and wind power are incorporated into hydropower plants such that all three forms of energy can act as a combination in order to deliver electricity to the grid. The power produced by the wind and PV plants is transferred to a control unit in a large-scale hydro–PV–wind hybrid system [43]. This, in turn, adjusts the hydropower generation to make up for the varying and irregular PV and wind electricity in short periods of time. It ensures that the total amount of energy provided to the power system satisfies the demands [44]. The advantages and disadvantages of this system are given in the Table 4.6.

4.4 Summary

In this chapter, different types of wind hybrid power technologies have been discussed. Every system has its advantages and disadvantages. Careful integration of these power generation systems can lessen their shortcomings and generate maximum electric power.

References

1. Makhsoos, A., Mousazadeh, H., Mohtasebi, S.S., Abdollahzadeh, M., Jafarbiglu, H., Omrani, E., Salmani, Y., Kiapey, A., Design, simulation and experimental evaluation of energy system for an unmanned surface vehicle. *Energy*, 148, 362, 2018.
2. Xu, X., Hu, W., Cao, D., Huang, Q., Chen, C., Chen, Z., Optimized sizing of a standalone PV-wind-hydropower station with pumped-storage installation hybrid energy system. *Renew. Energy*, 147, 1418, 2020.
3. Hou, P., Hu, W., Soltani, M., Chen, Z., Optimized placement of wind turbines in large-scale offshore wind farm using particle swarm optimization algorithm. *IEEE Trans. Sustain. Energy*, 6, 1272, 2015.

4. Silva, A.R. and Estanqueiro, A., From wind to hybrid: A contribution to the optimal design of utility-scale hybrid power plants. *Energies*, 15, 2560, 2022.
5. Zohuri, B., *Hybrid Energy Systems*, pp. 1–38, Springer Link, 2018, http://dx.doi.org/10.1007/978-3-319-70721-1_1.
6. Al-Thani, H., Koç, M., Isaifan, R.J., Bicer, Y., A review of the integrated renewable energy systems for sustainable urban mobility. *Sustain.*, 14, 10517, 2022.
7. Nsafon, B.E.K., Owolabi, A.B., Butu, H.M., Roh, J.W., Suh, D., Huh, J.S., Optimization and sustainability analysis of PV/wind/diesel hybrid energy system for decentralized energy generation. *Energy Strategy Rev.*, 32, 100570, 2020.
8. Mohapatra, A., Mohapatra, S., Bora, S.N., Hybrid renewable energy system for real-time power management techniques - a review, 2020, http://dx.doi.org/10.1109/isssc50941.2020.9358810.
9. Rathod, P., Mishra, S.K., Bhuyan, S.K., Renewable energy generation system connected to micro grid and analysis of energy management: A critical review. *Int. J. Power Electron. Drive Syst.*, 13, 470, 2022.
10. Ogbonnaya, C., Abeykoon, C., Damo, U.M., Turan, A., The current and emerging renewable energy technologies for power generation in Nigeria: A review. *Therm. Sci. Eng. Prog.*, 13, 100390, 2019.
11. McGowan, J.G., Manwell, J.F., Connors, S.R., Wind/diesel energy systems: Review of design options and recent developments. *Sol. Energy*, 41, 561, 1988.
12. Lamine, M., Belmokhtar, K., Agbossou, K., *New developments in renewable energy*, pp. 365, InTechopen, 2013, http://dx.doi.org/10.5772/52341.
13. Al-Janabi, Z.K. and Al-Janabi, S., An efficient predictor of renewable energy based on deep learning technique (DGBM) and multi-objectives optimization function, 2023, http://dx.doi.org/10.1109/iiccit55816.2022.10010380.
14. Bowen, A.J., Cowie, M., Zakay, N., The performance of a remote wind-diesel power system. *Renew. Energy*, 22, 429, 2001.
15. Shayeghi, H., Monfaredi, F., Dejamkhooy, A., Shafie-khah, M., Catalão, J.P.S., Assessing hybrid supercapacitor-battery energy storage for active power management in a wind-diesel system. *Int. J. Electr. Power Energy Syst.*, 125, 106391, 2021.
16. Ibrahim, H., Lefebvre, J., Methot, J.F., Deschenes, J.S., Numerical modeling wind-diesel hybrid system: Overview of the requirements, models and software tools, 2011, http://dx.doi.org/10.1109/epec.2011.6070204.
17. Das, U., Mandal, S., Bhattacharjee, S., Nandi, C., A review of different configuration of hybrid energy systems with case study analysis. *Int. J. Environ. Sustain. Dev.*, 21, 116, 2022.
18. Perera, A.T.D., Attalage, R.A., Perera, K.K.C.K., Dassanayake, V.P.C., Designing standalone hybrid energy systems minimizing initial investment, life cycle cost and pollutant emission. *Energy*, 54, 220, 2013.

19. Gupta, S.C., Kumar, Y., Agnihotri, G., Optimal sizing of solar-wind hybrid system, 2007, http://dx.doi.org/10.1049/ic:20070625.
20. Malik, M.Z., Zehra, K., Ali, I., Ubedullah, Ismail, M., Hussain, A., Kumar, V., Abid, M., Baloch, M.H., Solar-wind hybrid energy generation system, 2020, http://dx.doi.org/10.1109/inmic50486.2020.9318083.
21. Muhammad Lawan, S. and Azlan Wan Zainal Abidin, W., *Wind solar hybrid renewable energy system*, p. 95, InTechopen, 2020, http://dx.doi.org/10.5772/intechopen.85838.
22. Van, L.P., Chi, K.D., Duc, T.N., Review of hydrogen technologies based microgrid: Energy management systems, challenges and future recommendations. *Int. J. Hydrogen Energy*, 48, 14127, 2023.
23. Firtina-Ertis, I., Acar, C., Erturk, E., Optimal sizing design of an isolated stand-alone hybrid wind-hydrogen system for a zero-energy house. *Appl. Energy*, 274, 115244, 2020.
24. Hassan, Q., Abdulateef, A.M., Hafedh, S.A., Al-samari, A., Abdulateef, J., Sameen, A.Z., Salman, H.M., Al-Jiboory, A.K., Wieteska, S., Jaszczur, M., Renewable energy-to-green hydrogen: A review of main resources routes, processes and evaluation. *Int. J. Hydrogen Energy*, 48, 17383, 2023.
25. Apostolou, D. and Enevoldsen, P., The past, present and potential of hydrogen as a multifunctional storage application for wind power. *Renew. Sustain. Energy Rev.*, 112, 917, 2019.
26. Zhang, W., Maleki, A., Pourfayaz, F., Shadloo, M.S., An artificial intelligence approach to optimization of an off-grid hybrid wind/hydrogen system. *Int. J. Hydrogen Energy*, 46, 12725, 2021.
27. Li, Z., Liu, Y., Du, M., Cheng, Y., Shi, L., Modeling and multi-objective optimization of a stand-alone photovoltaic-wind turbine-hydrogen-battery hybrid energy system based on hysteresis band. *Int. J. Hydrogen Energy*, 48, 7959, 2023.
28. Zhang, W., Maleki, A., Rosen, M.A., Liu, J., Sizing a stand-alone solar-wind-hydrogen energy system using weather forecasting and a hybrid search optimization algorithm. *Energy Convers. Manage.*, 180, 609, 2019.
29. Chen, C., Liu, H., Xiao, Y., Zhu, F., Ding, L., Yang, F., Power generation scheduling for a hydro-wind-solar hybrid system: A systematic survey and prospect. *Energies*, 15, 8747, 2022.
30. Wang, Z., Fang, G., Wen, X., Tan, Q., Zhang, P., Liu, Z., Coordinated operation of conventional hydropower plants as hybrid pumped storage hydropower with wind and photovoltaic plants. *Energy Convers. Manage.*, 277, 116654, 2023.
31. Hemeida, A.M., El-Ahmar, M.H., El-Sayed, A.M., Hasanien, H.M., Alkhalaf, S., Esmail, M.F.C., Senjyu, T., Optimum design of hybrid wind/PV energy system for remote area. *Ain Shams Eng. J.*, 11, 11, 2020.
32. Jin, Y., Behrens, P., Tukker, A., Scherer, L., Water use of electricity technologies: A global meta-analysis. *Renew. Sustain. Energy Rev.*, 115, 109391, 2019.

33. Das, P., Das, B.K., Mustafi, N.N., Sakir, M.T., A review on pump-hydro storage for renewable and hybrid energy systems applications. *Energy Storage*, 3, e223, 2021.
34. Cui, J., Li, K., Sun, Y., Zou, Z., Ma, Y., Distributed energy storage system in wind power generation, 2011, http://dx.doi.org/10.1109/drpt.2011.5994140.
35. Ai, C., Zhang, L., Gao, W., Yang, G., Wu, D., Chen, L., Chen, W., Plummer, A., A review of energy storage technologies in hydraulic wind turbines. *Energy Convers. Manage.*, 264, 115584, 2022.
36. Thirunavukkarasu, M., Sawle, Y., Lala, H., A comprehensive review on optimization of hybrid renewable energy systems using various optimization techniques. *Renew. Sustain. Energy Rev.*, 176, 113192, 2023.
37. Mahmoud, M., Ramadan, M., Olabi, A.G., Pullen, K., Naher, S., A review of mechanical energy storage systems combined with wind and solar applications. *Energy Convers. Manage.*, 210, 112670, 2020.
38. Antunes Campos, R., Rafael do Nascimento, L., Rüther, R., The complementary nature between wind and photovoltaic generation in Brazil and the role of energy storage in utility-scale hybrid power plants. *Energy Convers. Manage.*, 221, 113160, 2020.
39. He, Y., Guo, S., Zhou, J., Wu, F., Huang, J., Pei, H., The quantitative techno-economic comparisons and multi-objective capacity optimization of wind-photovoltaic hybrid power system considering different energy storage technologies. *Energy Convers. Manage.*, 229, 113779, 2021.
40. Arabzadeh Saheli, M., Fazelpour, F., Soltani, N., Rosen, M.A., Performance analysis of a photovoltaic/wind/diesel hybrid power generation system for domestic utilization in Winnipeg, Manitoba, Canada. *Environ. Prog. Sustain. Energy*, 38, 548, 2019.
41. El-houari, H., Allouhi, A., Salameh, T., Kousksou, T., Jamil, A., El Amrani, B., Energy, economic, environment (3E) analysis of WT-PV-battery autonomous hybrid power plants in climatically varying regions. *Sustain. Energy Technol. Assess.*, 43, 100961, 2021.
42. Bhattacharjee, S. and Acharya, S., PV-wind hybrid power option for a low wind topography. *Energy Convers. Manage.*, 89, 942, 2015.
43. Peng, Z., Chen, X., Yao, L., Research status and future of hydro-related sustainable complementary multi-energy power generation. *Sustain. Futur.*, 3, 100042, 2021.
44. Wang, Z., Wen, X., Tan, Q., Fang, G., Lei, X., Wang, H., Yan, J., Potential assessment of large-scale hydro-photovoltaic-wind hybrid systems on a global scale. *Renew. Sustain. Energy Rev.*, 146, 111154, 2021.

5

Theories Based on Technological Advances for Wind Energy

Muhammad Azam, Laiba Zafar, Haq Nawaz Bhatti and Amina Khan*

Department of Chemistry, University of Agriculture, Faisalabad, Pakistan

Abstract

This chapter provides a comprehensive examination of wind energy theories based on technological advancements. Beginning with a comprehensive overview of the theoretical backdrop, including the fundamental principles of wind energy conversion and the aerodynamics of wind turbines, the text then delves into the practical applications of wind energy. The chapter then delves into wind turbine control systems, emphasizing their significance in optimizing performance and ensuring safe, efficient power generation. The chapter focuses primarily on theories based on technological advancements. These theories include aeroelasticity, wind turbine design, power control, wind farm layout, turbine placement, wake effect, grid integration, power quality, and stability, to name a few. Each theory is examined in depth, with a focus on its significance and the associated advances in the discipline. In addition, the chapter examines innovations in wind energy technologies, including advanced materials and manufacturing techniques, innovative turbine designs, enhanced control systems and sensors, and smart grid integration with energy storage. These developments are essential for enhancing the efficacy, dependability, and competitiveness of wind energy as a sustainable energy source.

Keywords: Wind energy, technological advances, theories, control systems, wind turbine design, grid integration, power quality, stability, future research directions

**Corresponding author*: aminakhan1649@gmail.com

Inamuddin, Tariq Altalhi and Mohammad Luqman (eds.) Wind Energy Storage and Conversion: From Basics to Utilities, (83–114) © 2024 Scrivener Publishing LLC

5.1 Introduction

Wind energy is currently one of the most rapidly expanding renewable energy sources around the globe. As we endeavor to reduce our dependence on fossil fuels and mitigate the effects of climate change, its significance has grown. Charles F. Brush constructed the first wind turbine to generate electricity in the United States in 1887 (Chiulan & Anton, 2021). Brush's machine featured a rotor with 144 blades measuring 17 m in diameter and a 12-kW generator. In Denmark, Poul La Cour conducted a fundamental wind turbine research between 1891 and 1908, oblivious to Brush's work. Based on his concepts, the design of aerodynamically efficient rotor blades advanced rapidly, and wind turbines provided approximately 3% of Denmark's electricity demand in 1918. Numerous benefits, such as its abundance, dependability, and affordability, make wind energy an attractive option for meeting our energy needs. Wind energy's significance in meeting our energy needs cannot be exaggerated. It is a pure, renewable, and abundant energy source that can help mitigate the effects of climate change and reduce greenhouse gas emissions. As the global population continues to increase, energy demand will continue to rise. Wind energy is a significant and swiftly expanding renewable energy source that has the potential to play a substantial role in meeting our energy needs (Shah & Solangi, 2019). The advancement of wind turbine and control technology over the preceding 20 years has been directly related to the widespread usage of wind energy. In the meantime, in 2021and 2022, an additional 102 and 77.6 GW of wind generating capacity was built, respectively (Figure 5.1).

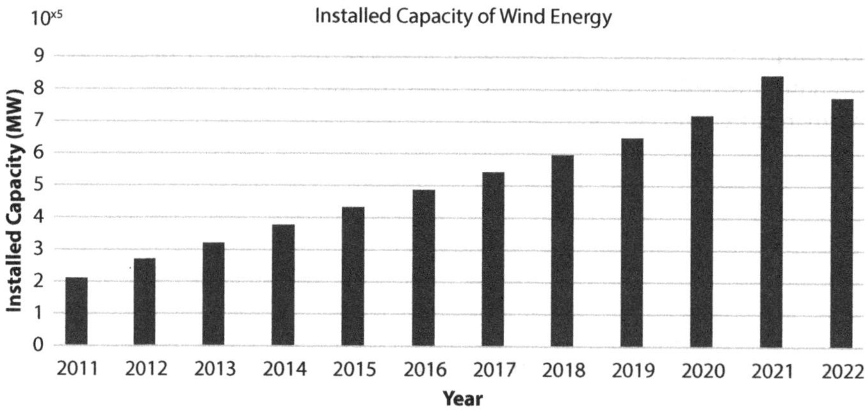

Figure 5.1 Installed capacity of wind power during recent years.

The overall capacity increased by 13.5% annually to exceed 922 GW (Jamal *et al.*, 2015).

Wind energy is based on converting the kinetic energy of wind into electricity using wind turbines. Wind turbines are designed to capture and convert the wind's energy into electricity. The components include a rotor, blades, shaft, generator, and tower (Figure 5.2). The airflow rotates the rotor, which, in turn, spins the blades (Wilberforce *et al.*, 2023). The generator, which produces electricity, is driven by the rotating blades. Wind energy has advanced significantly since its inception. Initially, wind turbines were basic, low-capacity machines that were primarily used to power remote homes and small businesses. However, significant technological advancements in wind energy systems have led to the development of wind turbines that are larger, more efficient, and more dependable over time. Today, wind energy is utilized on a larger scale, and wind farms are now a prevalent sight in many regions of the globe (McKenna *et al.*, 2022).

The advancements in wind energy technology have been driven by the need to increase the efficiency and dependability of wind turbines. This was made possible by the development of sophisticated materials, improved manufacturing techniques, innovative designs, and more efficient control systems (Nazir *et al.*, 2020). These developments have also reduced the cost of wind energy, making it competitive with other forms of energy. Horizontal-axis wind turbines (HAWTs) have undergone

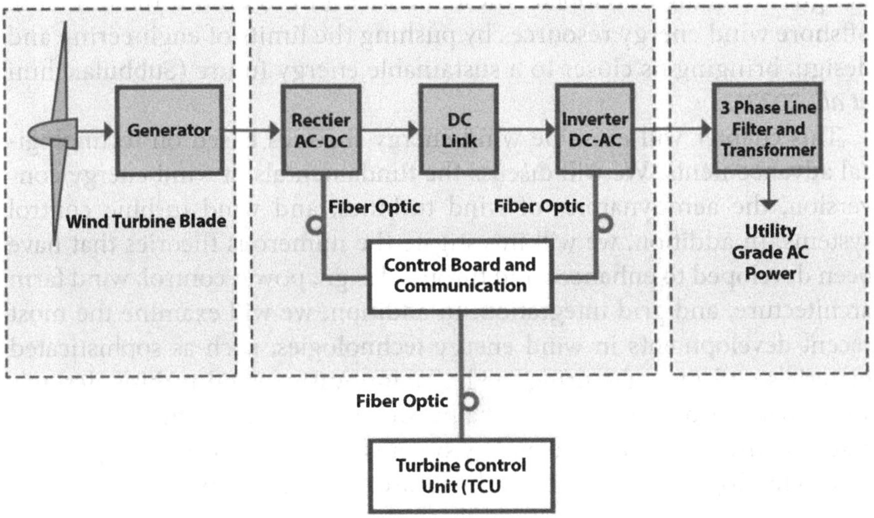

Figure 5.2 Block diagram of the components of a wind energy conversion system.

significant advancements, including new aerodynamic designs, sophisticated materials, and improved control mechanisms (Chaudhuri *et al.*, 2022). These developments have spawned theories such as the Betz limit, which explores the theoretical utmost limit of energy extraction from wind and offers insights for optimizing wind turbine performance. In addition, the concept of wake steering, which investigates the manipulation of wind turbine wakes to increase wind farm energy production, has gained traction (Houck, 2020).

In addition, the incorporation of cutting-edge technologies has expanded the scope of wind energy. The development of theories focusing on the intelligent administration and control of wind energy systems has been facilitated by the emergence of smart grids and digitalization. The application of machine learning algorithms and artificial intelligence techniques has facilitated the prediction and optimization of wind power generation, thereby contributing to the development of more effective operation and maintenance strategies. These theories aim to establish a symbiotic relationship between technological advances and wind energy production, thereby improving the economic viability and dependability of wind power. New hypotheses and studies have been sparked by the ongoing exploration of offshore wind energy. Offshore wind farms have the potential to capture stronger and more consistent winds, thereby increasing their energy production capacity. To overcome the distinct challenges posed by offshore environments, theories concerning floating wind turbines, dynamic cabling systems, and innovative foundation designs are being developed. These theories aim to unlock the immense offshore wind energy resources by pushing the limits of engineering and design, bringing us closer to a sustainable energy future (Subbulakshmi *et al.*, 2022).

This chapter will examine wind energy theories based on technological advancements. We will discuss the fundamentals of wind energy conversion, the aerodynamics of wind turbines, and wind turbine control systems. In addition, we will investigate the numerous theories that have been developed to enhance wind turbine design, power control, wind farm architecture, and grid integration. In addition, we will examine the most recent developments in wind energy technologies, such as sophisticated materials and manufacturing methods, innovative wind turbine designs, and smart grid integration. Finally, we will examine the prospective research directions for wind energy systems as well as the emerging trends and technologies that will shape the future of wind energy.

5.2 Theoretical Background

5.2.1 Basic Principles of Wind Energy Conversion

The conversion of wind energy to electricity is a complex process founded on several theoretical concepts. The fundamental principles of wind energy conversion stem from the physical and mechanical laws that govern the behavior of fluids and the motion of objects in a fluid medium. These principles constitute the basis of wind turbine aerodynamics and wind energy system design. Wind energy conversion is based on the transfer of kinetic energy from the wind to the rotor blades of a wind turbine. The rotation of the blades converts this kinetic energy into mechanical energy, which operates a generator to produce electrical energy. Wind speed and the area covered by the rotor blades determine the quantity of kinetic energy that can be extracted from the wind (Eltayesh *et al.*, 2021).

Wind turbines and converters are part of the wind energy conversion system (WECS). The mechanical energy of the wind is extracted by wind turbines, transformed into electrical energy by generators, and then transferred to the power grid or a battery bank *via* the converter system. The most popular generators used for converting wind energy to electricity at a variable speed are synchronous and doubly fed induction generators (DFIG). When utilizing induction generators, the electronic converter serves as a bridge between the rotor and stator, which are both connected to the network directly. The DFIG has long been the favored option for large, variable-speed WECS that are linked directly to the electrical network because the rotor circuit may independently adjust the amplitude and frequency of the generated voltage. Synchronous power generators are cut off from the grid that distributes electricity. They are thus included into systems that make use of full-scale converters (Carrasco *et al.*, 2006).

The fundamental principles of fluid mechanics, which characterize the behavior of fluids in motion, must be considered to comprehend wind energy conversion. Fluid mechanics is the branch of physics that examines the behavior of fluids, such as liquids and gases, under different conditions. The wind is a fluid medium that interacts with a turbine's rotor blades, so the principles of fluid mechanics are fundamental to the design and operation of wind turbines. Bernoulli's principle is one of the key fluid mechanics principles pertinent to wind energy conversion. This principle asserts that a fluid's pressure will decrease as its velocity increases. The rotor blades of wind turbines are designed to generate a pressure difference between the top and bottom surfaces of the blade, causing the blade to rotate. Ultimately, this rotation is what causes the generator to produce electricity (Mu *et al.*, 2022).

The concept of lift is an additional essential fluid mechanics principle in the context of wind energy conversion. When a fluid travels over a curved surface, such as the rotor blades of a wind turbine, lift is generated. The configuration of the rotor blades is meticulously designed to generate lift, which aids in wind energy extraction. In the discipline of fluid mechanics, the theoretical principles governing the behavior of fluids and the motion of objects in a fluid medium are well established. The fundamental equations of fluid mechanics, such as the Navier–Stokes and Bernoulli equations, describe the flow of fluids and the forces operating on objects in a fluid medium. These equations are the premise for wind turbine aerodynamics and the design of wind energy systems. In general, the fundamental principles of wind energy conversion derive from the laws of physics and mechanics, specifically the principles of fluid mechanics. These principles serve as the basis for the design and operation of wind turbines as well as the efficient conversion of wind energy to electricity (Nishino & Dunstan, 2020).

5.2.2 Aerodynamics of Wind Turbines

The aerodynamics of wind turbines is a complex discipline that examines the flow of air over a wind turbine's rotor blades. The design of the rotor blades is essential to the efficacy and performance of the wind turbine because it determines how much energy can be extracted from the wind. The aerodynamics of wind turbines are founded on the fluid mechanics principles that describe the behavior of moving fluids (Azam *et al.*, 2021). The airflow over a wind turbine's rotor blades is governed by the Navier–Stokes equations, which are the fundamental equations of fluid mechanics. These equations characterize fluid motion and the forces acting on objects in a fluid medium. The concept of lift, which is the force generated when a fluid travels over a curved surface, is one of the fundamental concepts in the aerodynamics of wind turbines. A wind turbine's rotor blades are meticulously designed to generate lift, which serves to extract energy from the wind. The shape of the rotor blades is also crucial because it affects the quantity of lift produced by the blades.

The concept of drag is another essential concept in the aerodynamics of wind turbines. The force that opposes the motion and is generated when a fluid flows over a surface is known as drag (Figure 5.3). In the case of wind turbines, drag is produced by the air's interaction with the rotor blades' surfaces. To optimize the efficiency of wind turbines, it is crucial to minimize

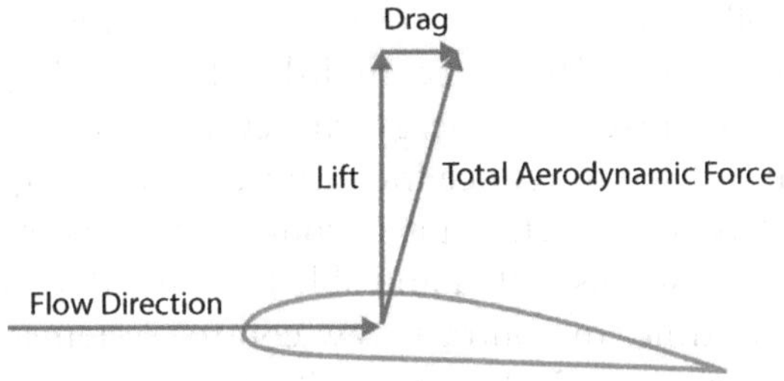

Figure 5.3 The concept of drag.

drag. The aerodynamics of wind turbines is an active area of study, with ongoing endeavors to improve the performance and efficiency of wind turbines. Researchers are investigating novel rotor blade designs, such as advanced airfoil shapes and active control systems. They are also investigating the use of multiple wind turbines in wind farms and the behavior of wind turbines in various wind conditions. There have been numerous investigations on the aerodynamics of wind turbines. Using computational fluid dynamics simulations, Lu and colleagues analyzed the flow of air over the rotor blades of a wind turbine. The efficacy of the wind turbine was found to be affected by variables such as wind speed, blade angle, and blade shape (Sun *et al.*, 2020).

Rohlfs and colleagues (2019) investigated the effect of wind shear on the efficacy of wind turbines in a separate study. Wind shear is the variation in wind speed and direction with altitude, which can influence wind turbine behavior. The researchers discovered that wind shear can have a significant effect on the efficiency of wind turbines and recommended the use of active control systems to mitigate this effect. Overall, wind turbine aerodynamics is a complex discipline based on the principles of fluid mechanics. The design of the rotor blades is essential to the efficacy and performance of wind turbines, and ongoing research focuses on enhancing their aerodynamic performance (Xie & Aly, 2020).

5.2.3 Control Systems for Wind Turbines

Wind turbine operation is fundamentally dependent on control systems. These systems are responsible for regulating the turbine's speed and power output, as well as ensuring its safety and the safety of its surroundings. Wind turbines utilize two primary categories of control systems: pitch control systems and yaw control systems (Sitharthan *et al.*, 2019).

The responsibility of pitch control systems is to regulate the angle of attack of the rotor blades. This is accomplished by adjusting the pitch angle of the blades to maximize wind energy extraction. In high gusts, pitch control systems can be used to reduce the turbine's power output and prevent blade damage. Passive or active pitch control systems exist, with active systems being more sophisticated and able to respond to changes in wind conditions in real time. In contrast, yaw control systems are responsible for aligning the turbine with the airflow. This is accomplished by rotating the nacelle (the turbine's generator enclosure) around a vertical axis. Wind turbines are most effective when they are facing directly into the wind, so yaw control systems are crucial for maximizing wind energy extraction. By orienting the turbine to minimize wind loading on the blades, yaw control systems can also be used to prevent damage to the turbine in high winds (Menezes & Araújo, 2023).

With the development of enhanced sensing and control technologies, wind turbine control systems have become progressively more sophisticated in recent years. Modern wind turbines, for instance, are outfitted with sensors that can measure wind speed, direction, and turbulence as well as the position and movement of the rotor blades. The control system then uses this data to optimize the turbine's performance and ensure its safe operation. Numerous studies have been conducted on the design and efficacy of wind turbine control systems. Kühn and associates (2019), for instance, investigated the application of model predictive control (MPC) for pitch control in wind turbines. MPC is an advanced control strategy that employs mathematical models of the system to anticipate future behavior and optimize control inputs. The researchers discovered that MPC can enhance the performance of pitch control systems, notably in conditions of turbulent wind (Vali *et al.*, 2019).

Hannan and colleagues (2023) investigated the use of fuzzy logic controllers (FLCs) for yaw control in wind turbines in a separate study (Hannan *et al.*, 2023). FLC is a type of control strategy that makes decisions based on imprecise or ambiguous information using fuzzy logic. The researchers discovered that the FLC-based control system improved the wind turbine's yaw control performance, particularly in situations where the wind direction changed swiftly. Control systems are an essential component of wind turbines, and ongoing research focuses on enhancing their design and performance. Future wind turbine control systems are anticipated to be more sophisticated and efficient due to developments in sensing and control technologies. In the sphere of wind energy, wind turbine control systems have been extensively studied and are well established (R. Kumar *et al.*, 2021).

5.3 Theories Based on Technological Advances

Numerous hypotheses aiming to improve the performance and efficiency of wind turbines have been developed as a result of technological advances in wind energy. Based on a combination of engineering principles, mathematical modeling, and experimental data, these theories are continuously evolving as new technologies and data emerge. To improve the operation of wind turbines and energy production in the modern era, three fundamental areas of research are proposed (Torregrosa *et al.*, 2021), namely:

- Aeroelasticity
- Computational fluid dynamics
- Artificial intelligence

A. Aeroelasticity
The study of the interaction between the aerodynamic forces operating on a structure and its structural dynamics is aeroelasticity. Aeroelasticity in the context of wind energy is concerned with the behavior of wind turbines under varying wind conditions and how this behavior impacts the performance and safety of the turbine. Several aeroelasticity-based theories have been created to enhance the efficacy and efficiency of wind turbines—for instance, the "smart rotor" theory proposes the use of active control systems to modify the shape of the rotor blades based on wind conditions and other variables in real time. This theory seeks to maximize wind energy extraction while reducing the turbine's and its components' loads (Maza *et al.*, 2023).

B. Computational fluid dynamics
Computational fluid dynamics (CFD) is an engineering discipline that simulates the behavior of fluids, such as air and water, using numerical methods. CFD is used to model the aerodynamics of wind turbines and predict their efficacy under varying wind conditions in the context of wind energy. Several CFD-based theories have been devised to enhance the design and functionality of wind turbines—for instance, the "blade element momentum" theory simulates the performance of wind turbine blades using a combination of CFD and mathematical modeling. Using variables such as wind speed, turbulence, and blade geometry, this theory seeks to optimize the design of the blades (Bhatti *et al.*, 2020).

C. Artificial intelligence
Artificial intelligence (AI) is a rapidly developing discipline concerned with the creation of computer systems that can perform tasks that typically

require human intelligence, such as learning, reasoning, and decision-making. AI is used to develop sophisticated control systems for wind turbines and optimize their performance and maintenance in the context of wind energy. Several wind energy theories founded on AI have been developed. The "machine learning" theory, for instance, proposes the use of machine learning algorithms to analyze large amounts of data from wind turbines and identify patterns that can be used to enhance their performance and maintenance. This theory seeks to improve the overall performance and efficiency of wind turbines while decreasing their downtime and maintenance costs (X. Zhao *et al.*, 2019).

Wind turbine performance and efficiency have been significantly enhanced by the development of theories founded on technological advancements. As new technologies and data become available, these theories are continuously evolving and are expected to continue driving innovation in the wind energy sector.

5.3.1 Wind Turbine Design Theory

The theory of wind turbine design refers to the principles and methodologies used to design and optimize wind turbines for efficient and dependable operation. This comprises, among other components, the design of the rotor blades, tower, nacelle, and control systems. The most important aspect of the wind turbine design theory is the optimization of rotor blade size and shape. This involves striking a balance between the requirements for high energy capture, structural integrity, and low noise emissions. Optimizing the blade twist angle, chord length, and airfoil shape is one of the commonly used design strategies. Choosing the optimal control systems to optimize the operation of the turbine is a further consideration. This includes the use of pitch control to adjust the angle of the blades, yaw control to adjust the direction of the turbine, and power control to modify the output of the generator. In addition, advanced control systems can be used to optimize the rotor speed and blade pitch angle in real time based on wind conditions and other variables. By optimizing the design of wind turbines, we can increase their energy output, decrease their environmental impact, and increase their durability and dependability (Hart *et al.*, 2020).

The development of new wind turbine models, such as the Vestas V164-9.5 MW turbine, which features a 164-m rotor diameter and sophisticated control systems to optimize performance and reliability, is an example of the application of the wind turbine design theory (Astolfi *et al.*, 2021). The GE Haliade-X turbine has a 12-MW capacity and a rotor diameter of 220 m as well as advanced control systems and a modular design for simple transportation and installation (O'Neill, 2021).

5.3.1.1 *Rotor Blade Design Theory*

Rotor blade design theory refers to the principles and methodologies used to design and optimize wind turbine rotor blades for efficient and reliable operation. The rotor blades are a crucial component of the turbine, as they are responsible for converting wind energy into rotational energy (Alom & Saha, 2019). These are the main characteristics of this theory:

- The selection of the appropriate airfoil shape.
- The optimization of the blade twist angle and chord length.

The lift and drag forces generated by the blade are affected by the shape of the airfoil, which, in turn, impacts its overall energy capture and efficiency. Numerous airfoil shapes, including symmetric and asymmetric shapes as well as specialized shapes optimized for specific wind conditions, have been devised and studied for wind turbine blades. The twist angle is the angle between the blade's base and tip, while the chord length is the blade's width at each position along its length. The twist angle and chord length influence the blade's angle of attack and, consequently, its overall lift and drag forces. Various optimization techniques, including numerical simulations, experimental testing, and artificial intelligence algorithms, have been devised for rotor blade design (J. Zhao *et al.*, 2022).

The materials employed in the design of rotor blades are also crucial to their performance and dependability. Due to their high strength-to-weight ratio and resistance to fatigue and corrosion, composite materials, such as fiber glass and carbon fiber, are widely used in modern wind turbine blades. Other materials, such as wood and metal, have been utilized historically but are less common in modern wind turbines due to their inferior performance. Examples of the application of rotor blade design theory include the development of new wind turbine models, such as the Siemens Gamesa SG 14-222 DD turbine with 108-m-long blades and a 14-MW capacity and advanced blade design features to improve performance and reliability (Hassan *et al.*, 2020). By optimizing the design of rotor blades, we can increase the energy output of wind energy systems, reduce their environmental impact, and improve their durability and dependability.

The rotor blades can be controlled using a system called the smart rotor control system. It refers to the use of advanced sensors, control systems, and algorithms to actively modify the aerodynamic performance of wind turbine rotor blades in response to changing wind conditions in real time. By optimizing the aerodynamic performance of the rotor blades, intelligent rotor control can enhance energy capture efficiency, reduce stresses on the blades, and boost the wind turbine's overall power output. Smart rotor control systems consist of sensors that measure wind speed, direction, and other variables as well as controllers that alter the pitch angle and twist of the rotor blades based on the data collected by the sensors. Using sophisticated algorithms, the controllers determine the optimal settings for the rotor blades based on the current wind conditions and make the necessary adjustments. By modifying the rotor blade settings to minimize blade–vortex interactions and other noise sources, smart rotor control can also be used to reduce noise emissions from wind turbines. Wind turbine operators can strengthen their relationships with neighboring communities and reduce the environmental impact of wind energy systems by reducing noise emissions (Hu *et al.*, 2023).

Individual pitch control (IPC) systems, which adjust the pitch angle of each blade based on local wind conditions rather than adjusting all blades as a single entity, are an example of smart rotor control in action. By reducing the negative effects of wind shear and turbulence, which can cause uneven loading on the rotor blades and reduce their overall efficiency, IPC systems can increase the energy capture efficiency of the turbine.

5.3.1.2 Aerodynamic Design Theory

The design of wind turbine blades must balance competing design objectives, such as maximizing energy capture efficiency while minimizing loads and pollution. The study and application of fluid dynamics principles in the design of wind turbine blades are referred to as aerodynamic design theory. The objective of aerodynamic design theory is to optimize the blades' shape and performance to maximize energy capture efficiency, minimize loads and pollution, and ensure the turbine's safe operation. It also incorporates the study of the boundary layer, which is the thin layer of air that flows over the wind turbine blade's surface. The behavior of the boundary layer can have a significant effect on the blade's aerodynamic performance and efficiency. Researchers and engineers have developed new techniques and instruments for modeling, simulating, and testing wind turbine blades, resulting in significant advances in aerodynamic design theory over the past few years. These developments have resulted in the creation of new blade designs that are more efficient, dependable, and cost-effective than their predecessors.

Blade element momentum (BEM) theory and computational fluid dynamics (CFD) simulations can assist in optimizing the blade geometry and performance (Khan *et al.*, 2021). The aerodynamic efficacy of the blades, including airfoil shapes, blade angles, and turbulence effects, can be determined through CFD simulations. By utilizing CFD simulations to optimize the design of wind turbine blades, researchers and engineers can increase the turbine's energy capture efficiency and decrease stresses and noise (Kwok & Hu, 2023).

Utilizing advanced materials and manufacturing processes to create wind turbine blades that are lighter, stronger, and more aerodynamic is an additional significant area of research in aerodynamic design theory—for instance, researchers are investigating the use of carbon fiber composites and other advanced materials to reduce the weight of the blades without compromising their structural integrity. In addition, researchers are exploring the use of 3D printing and other additive manufacturing techniques to produce blades with more complex geometries and enhanced aerodynamic performance (Xu *et al.*, 2021). In addition to advanced materials and fabrication methods, researchers are also investigating new blade designs that integrate active flow control systems. These systems use sensors and actuators to manipulate the airflow over the blade's surface, thereby enhancing its aerodynamic performance and minimizing disturbance (Guoqiang *et al.*, 2019).

The study of blade–tower interaction, which refers to the complex aerodynamic interactions between wind turbine blades and the structure, is also a component of aerodynamic design theory. Researchers are developing new techniques to model and simulate blade–tower interaction to facilitate the design of wind turbines that are stable and secure to operate. Utilizing serrated trailing edges on wind turbine blades is an illustration of a recent aerodynamic design advancement. Doolan *et al.* (2002) report that serrated trailing edges can minimize blade noise while also improving their aerodynamic performance (Jiang *et al.*, 2021). Another example is the use of curved, twisted, or tapered blade designs, which can increase the turbine's energy capture efficiency and reduce pollution and loads. The aerodynamic design is a crucial area of wind energy research, with the potential to considerably enhance the efficiency, dependability, and environmental impact of wind turbines. Researchers and engineers can create new blade designs that are more efficient, dependable, and cost-effective than their predecessors due to advancements in computational modeling, advanced materials, and manufacturing techniques.

5.3.2 Power Control Theory

An essential area of wind energy research, power control theory focuses on techniques for regulating the power output of wind turbines. This is essential for maximizing energy production and assuring stable integration with the electrical grid. There are numerous power control strategies, such as pitch control, torque control, and generator control. The most prevalent method for controlling the amount of wind energy captured is pitch control, which entails adjusting the pitch angle of the blades. It can adapt to variations in wind speed and direction to maximize energy production while minimizing turbine loads (Chavero-Navarrete *et al.*, 2019).

Another technique for managing power is torque control, which adjusts the generator's torque to manage the power output of the turbine. This strategy is especially beneficial in high-wind conditions, where pitch control may be ineffective. The third method is generator control, which modulates the electrical output of the generator to meet the needs of the grid. This method is beneficial for wind farms connected to the grid because it helps maintain grid stability and prevents power quality problems.

Recent research in power control theory has centered on the development of innovative techniques to increase the efficacy and dependability of wind turbines. Using machine learning algorithms to optimize real-time power control based on inputs such as wind speed, direction, and turbulence is one approach. The development of advanced control systems that take into consideration the dynamic behavior of wind turbines and their interaction with the grid is another strategy. These methods can enhance the stability and efficacy of turbines under various operating conditions. In addition, significant research has been conducted on the integration of energy storage systems with wind turbines to enhance power control. Energy storage systems can help smooth out fluctuations in power output caused by varying wind conditions and provide reserve power during grid failures. The incorporation of energy storage with wind turbines has the potential to increase the reliability and dispatchability of wind energy, thereby increasing its value to the grid (Lai & Teh, 2022).

As wind energy continues to expand as a significant source of renewable energy, it is crucial to enhance the efficacy and dependability of wind turbines. The theory of power control is essential to this endeavor and will continue to be an active area of study in the coming years. By developing sophisticated power control techniques, we can increase the amount of wind energy that can be captured and guarantee the grid's stable and dependable power supply.

5.3.2.1 Maximum Power Point Tracking Theory

Maximum Power Point Tracking (MPPT) is a technique used to optimize the wind turbine's output power in wind energy systems. MPPT operates the wind turbine at its maximum power point (MPP) by adjusting the turbine rotor speed and blade pitch angle to extract the maximum quantity of energy from the wind (Lai & Teh, 2022)(G. B. A. Kumar, 2022). The maximum power point (MPP) is the moment at which the turbine's power output is at its maximum for a given wind speed and is dependent on several variables, including wind speed, blade pitch angle, and rotor speed. MPPT controllers utilize various algorithms to continuously modify the turbine rotor speed and blade pitch angle to track the MPP. These algorithms include perturb and observe (P&O), incremental conductance (INC), and hill climbing (Kishor *et al.*, 2021). Due to its simplicity and dependability, the P&O algorithm is the algorithm most commonly used in wind energy systems. It is founded on the following equations:

$$V(t) = Vpv(t) + I(t)R \tag{5.1}$$

$$P(t) = V(t)I(t) \tag{5.2}$$

$$dP(t)/dI(t) = V(t) - I(t)R \tag{5.3}$$

where:

V(t) = the voltage of the wind turbine,
Vpv(t) = the voltage of the photovoltaic panel,
I(t) = the current of the wind turbine,
R = the resistance of the load, and
P(t) = the power output of the turbine.

Figure 5.4 depicts the MPPT controller as consisting of a wind turbine, a DC–DC converter, and an MPPT algorithm. The wind turbine generates AC power, which is converted by the converter into DC electricity. The MPPT algorithm modifies the rotor speed and pitch angle of the blades to keep the turbine operating at the MPP. Using MPPT techniques has been shown to increase wind turbine efficiency by up to 30% and reduce energy production costs. It also facilitates the incorporation of wind energy into the power grid by providing a consistent and constant power output. Variability in wind speed and direction is one of the greatest obstacles to implementing MPPT techniques. The control algorithms must be sufficiently robust to accommodate these variations and maintain turbine operation at the MPP. Numerous studies have been conducted to develop advanced MPPT algorithms capable of handling these variations and enhancing the performance of wind turbines.

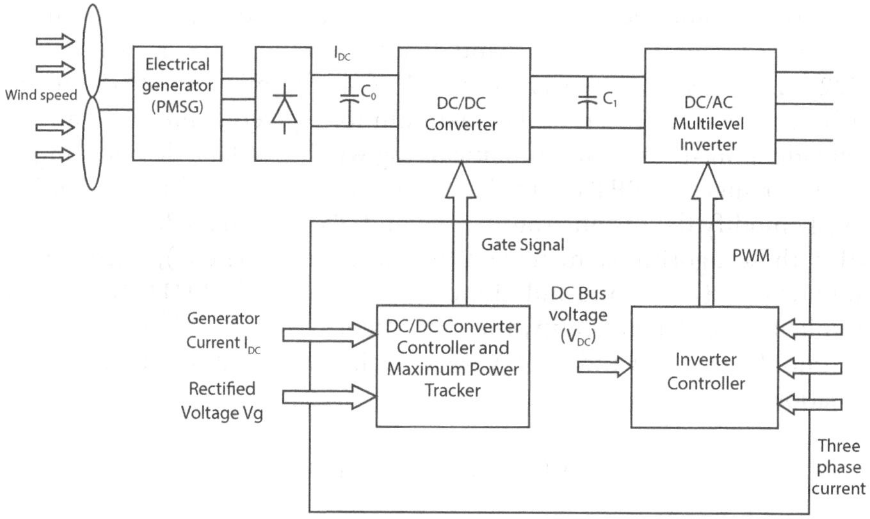

Figure 5.4 Block diagram of an MPPT controller for a wind turbine.

Xuewei Sun *et al.*, for example, proposed an advanced MPPT algorithm based on the fractional-order sliding mode control theory that can manage nonlinearity and uncertainty in wind energy systems (Vakacharla *et al.*, 2020). It was demonstrated that the algorithm is more robust and effective than the traditional P&O algorithm. Wei Qiao *et al.* proposed in a separate study an MPPT algorithm based on model predictive control theory that can estimate the MPP based on the predicted wind speed and direction. The algorithm has been shown to increase wind turbine efficacy by up to 5%. MPPT techniques are crucial for optimizing wind turbine output power and integrating wind energy into the power grid. To enhance the efficiency and robustness of MPPT controllers, advanced control algorithms based on modern control theory have been proposed.

5.3.2.2 *Load Control Theory*

In wind energy systems, load control theory concentrates on regulating and managing the power output to meet the needs of the connected electrical load or the grid (Liu *et al.*, 2023). It involves adjusting the power output of wind turbines to sustain grid stability while ensuring a stable and reliable power supply. Load control theory is essential for optimizing energy production, preventing power quality problems, and facilitating the incorporation of wind energy into the existing power infrastructure. Implementing sophisticated control strategies to modulate the power output of wind turbines is one of the most significant aspects of load control theory. These strategies ensure that the power generated by the turbines corresponds to the grid's or load's needs. Load control techniques can be categorized as curtailment control, reactive power control, and frequency control, among others.

A. Curtailment control
Curtailment control is a method for regulating the power output of wind turbines by curtailing or limiting the generation of electricity when essential. It is typically employed when the electrical discharge demand is less than the available wind power or when grid constraints exist. Controlling curtailment prevents grid instability, congestion, and excessive power injection. Advanced control algorithms and real-time monitoring systems are used to determine the level of curtailment based on grid conditions and to guarantee optimal power flow.

B. Reactive power control
Another aspect of load control theory, reactive power control, concentrates on regulating the reactive power output of wind turbines. Reactive power is necessary to maintain voltage stability and grid balance. Wind turbines can support reactive power by altering the injection or absorption of reactive power. Monitoring the grid voltage and power factor and dynamically adjusting the reactive power output of the turbines to maintain the intended grid voltage and power factor levels constitute reactive power control techniques.

C. Frequency control
Frequency control is essential for sustaining grid stability, particularly in power systems with a high wind energy penetration. Wind turbines with sophisticated load control systems can actively contribute to frequency regulation. These systems monitor the grid frequency and adjust the power output of the turbines to help maintain an acceptable grid frequency. Wind turbines improve grid reliability and reduce reliance on conventional power facilities for frequency control by providing frequency support.

5.3.3 Wind Farm Layout Theory

Wind farm layout theory is a crucial component of wind energy project development. It involves optimizing the placement of wind turbines and the interconnection of turbines within a wind farm to accomplish maximum energy production at the lowest possible cost and environmental impact (Lerch *et al.*, 2019). The distance between wind turbines is a crucial factor in wind farm layout theory, as it can considerably affect the power output of the wind farm. Studies have demonstrated that the optimal distance between turbines can vary based on wind speed and topography. In regions with high wind speeds and flat terrain, for instance, turbines can be placed closer together, whereas in regions with low wind speeds and complex terrain, turbines should be placed further apart to prevent turbulence effects. The choice of wind turbine model and hub height is an additional essential aspect of wind farm layout theory. Wind speed and site-specific conditions can influence the optimal turbine model and hub height. In areas with moderate wind speeds, for instance, it may be more efficient to use larger turbines with higher hub heights to capture more wind energy.

Recent advancements in wind farm layout theory have led to the development of novel optimization strategies for wind farm layouts. Utilizing advanced simulation models to predict the power output of wind farms under various conditions is one such method. Using variables such as wind speed, turbulence, and terrain, these models can determine the optimal turbine placement and interconnection for a given site. Utilizing optimization algorithms and machine learning techniques is an additional approach to wind farm layout theory. These methods can be used to analyze large quantities of data and determine the optimal wind farm layout based on a variety of variables, such as wind speed, terrain, and environmental impact. By optimizing the placement and interconnection of wind turbines, the energy output of the wind farm can be maximized while its environmental impact is minimized (Hou *et al.*, 2019).

5.3.3.1 Turbine Placement Theory

The placement of turbines is an essential aspect of wind farm design and optimization. It entails determining the optimal placement of wind turbines within a wind farm to maximize energy production, minimize wake effects, and guarantee efficient operation. Numerous factors, including wind resource characteristics, site topography, environmental considerations, and turbine interactions, influence the arrangement of turbines within a wind farm. The objective is to position the turbines to capture the maximum amount of available wind energy while minimizing wake effects, which can reduce power output and increase structural stresses on turbines downstream. The site's wind flow characteristics are a crucial consideration in turbine placement theory. For optimal turbine placement, it is essential to identify areas with high wind speeds and minimal turbulence intensity. Wind resource assessments, which involve the measurement and analysis of wind data at various heights and locations, play a crucial role in determining the optimal installation locations for wind turbines. The topography of the site also affects turbine placement. Hills, ridges, valleys, and bodies of water can have a substantial effect on wind flow patterns and turbulence. The placement of wind turbines should take the terrain into account and ensure that turbines are positioned to capture the utmost available wind resource while avoiding areas of high turbulence or flow separation (Caglayan *et al.*, 2019).

Consideration of wake effects is an additional factor in turbine placement theory. When a wind turbine extracts energy from the wind, it generates a wake of slower-moving air downstream, which can hinder the performance of turbines located downstream. Proper turbine spacing and layout are essential to minimize disturbance effects and optimize wind farm energy production. Using advanced wake modeling and simulation techniques, wake interactions are analyzed and turbine placement is optimized accordingly. In turbine placement theory, environmental factors such as wildlife habitats, avian migration patterns, and visual impact also play a role. It is essential to achieve a balance between maximizing energy production and minimizing potential negative environmental effects. Various techniques, such as computational models, optimization algorithms, and machine learning strategies, are utilized to determine the optimal turbine placement. These methods consider multiple variables, such as wind data, site characteristics, discharge effects, and constraints, to determine the most efficient and effective turbine layout for a particular wind farm project. Wind farm developers can design layouts that optimize energy production, minimize wake effects, and address environmental concerns by employing turbine placement theory (Hou *et al.*, 2019).

5.3.3.2 Wake Effect Theory

Wake effect theory is a foundational concept in wind energy that describes the phenomenon in which downstream wind turbines in a wind farm experience reduced wind speeds and altered wind flow patterns as a result of the wake created by upstream turbines. Understanding and managing wake effects is essential to optimize wind farm performance and efficiency. When a wind turbine extracts energy from the wind, a turbulent trail is produced downstream. This wake consists of air that is traveling more slowly and is more turbulent than the unaffected wind flow. Consequently, the downstream turbines experience decreased wind velocities, increased turbulence, and altered wind directions. The wake effect can have several deleterious effects on the performance of turbines downstream. Initially, the decreased wind speed reduces the available kinetic energy that can be converted into electrical power, resulting in a decrease in power output. Secondly, the increased turbulence can impose additional stresses on the turbine blades, diminishing their structural integrity and service life. Moreover, the altered wind direction within the aftermath can result in suboptimal inflow angles for the turbines downstream, further diminishing their performance (Famoso *et al.*, 2020). Diverse strategies and technologies are utilized to mitigate the negative effects of wake effects, namely:

- Turbine location: Appropriate turbine spacing and layout are essential to minimize disturbance effects. Increasing the distance between turbines allows for greater wind speed recovery and reduced wake turbulence.
- Wake modeling and simulation: Using advanced computational models and simulations, wake effects in wind farms are predicted and analyzed. To accurately simulate wake behavior, these models evaluate wind conditions, turbine characteristics, terrain, and atmospheric stability. This data is then used to optimize the placement and layout of turbines.
- Wake steering: Wake steering is the process of modifying the yaw angle of upstream turbines to divert the wake away from downstream turbines, thereby reducing the impact on their performance. Advanced control strategies and real-time monitoring of wind conditions can achieve this technique.
- Innovative turbine design: Innovative turbine designs with enhanced aerodynamics and reduced wake generation can aid in minimizing wake effects. Turbines with larger rotor diameters and optimized blade configurations, for instance, can reduce wake turbulence and increase power output.
- Curtailment strategies: In some instances, the power output of specific turbines or individual blades can be reduced to reduce disturbance effects and improve the overall performance of a wind farm. This strategy adjusts turbine operation based on real-time wind conditions and power demand.

Continual efforts are made to refine the wake effect theory and develop cutting-edge optimization techniques for wind farms. The negative effects of wake effects can be mitigated through improved turbine placement, advanced modeling, wake steering, and inventive turbine designs, resulting in increased energy production and enhanced wind farm efficiency.

5.3.4 Grid Integration Theory

Grid integration theory refers to the process of connecting wind farms to the electrical grid to supply energy to consumers. Integration of wind energy into the power grid necessitates specific measures to ensure the system's dependability, stability, and efficiency. The intermittent and variable nature of wind power makes it difficult to predict and manage the power supply, which is the primary obstacle to integrating wind energy into the grid. To overcome this difficulty, sophisticated control and management systems are necessary to assure a reliable and stable power supply. This theory describes the use of advanced power electronics to assure the compatibility of wind-generated electricity with the electrical grid. This requires the use of power conditioning systems, such as inverters, which convert the DC power generated by wind turbines into grid-compatible AC power (Sayahi *et al.*, 2020).

Grid integration theory comprises several essential components, namely:

- Grid stability and power quality: Wind energy is intermittent and variable due to variations in wind speed and direction. Despite these fluctuations, grid integration theory emphasizes maintaining grid stability and power quality. Advanced control and monitoring systems are utilized to manage power flow, regulate voltage and frequency, and mitigate potential grid disturbances induced by the injection of wind power.
- Grid connection and infrastructure: This theory examines the operational and technical factors involved in connecting wind farms to the grid. It entails selecting suitable grid connection points, designing efficient transmission and distribution infrastructure, and ensuring compliance with grid codes and regulations. A proper grid connection facilitates efficient power transfer and reduces transmission losses.
- Energy forecasting and scheduling: For optimal grid integration, accurate forecasting of wind power generation is essential. Grid operators can make informed decisions regarding power scheduling, reserve capacity, and grid balancing by using advanced forecasting models and algorithms to anticipate wind power output in advance. Effective energy scheduling ensures that consumers have a reliable and stable power supply.

- Ancillary services and grid support: Wind farms can provide valuable ancillary services to support the operation of the grid. The theory of grid integration includes the provision of services like frequency regulation, voltage control, and reactive power support. Wind farms can be outfitted with sophisticated control capabilities to react to grid signals and contribute to grid stability during system disturbances.
- Grid codes and standards: Grid integration theory requires comprehension and adherence to grid codes and standards established by grid operators and regulatory agencies. These codes define technical requirements, performance criteria, and operational guidelines for wind farms to ensure that their grid integration is safe and reliable. Compliance with grid codes facilitates grid connection and reduces grid instability risk.
- Market and economic considerations: The theory of grid integration also addresses market mechanisms, economic incentives, and policies that promote the efficient integration of wind power. It examines strategies such as feed-in tariffs, power purchase agreements, and market-based mechanisms for promoting renewable energy generation and encouraging grid-friendly operation.
- Smart grid technologies: Wind power integration is aided by the development of smart grid technologies. Smart grids permit wind farms, grid operators, and consumers to communicate, exchange data, and exercise control in real time. These technologies improve grid flexibility, enable demand response programs, and facilitate the integration of renewable energy sources in an efficient manner.

5.3.4.1 Power Quality Theory

Power quality theory is the study of the characteristics of electrical power and the methods used to ensure that the power supplied satisfies the required quality standards. Power quality is an essential aspect of wind energy systems because it ensures the reliable and efficient operation of grid-connected electrical equipment. Due to the intermittent and variable nature of wind power generation, wind energy systems may encounter power quality issues. Among these problems are voltage fluctuations, harmonics, flicker, and voltage imbalance. These disturbances can have an effect on the performance and longevity of electrical equipment, lead to malfunctions, and increase maintenance costs (Kharrazi *et al.*, 2020).

Several theories and methods are utilized to address power quality concerns, namely:

- Voltage regulation: Voltage regulation is necessary to maintain a stable and dependable power supply. Wind turbines are equipped with power electronics systems that regulate and monitor the output voltage to meet grid requirements. Adjusting the power output of wind turbines and maintaining a stable grid voltage are made possible using sophisticated control algorithms.
- Harmonic mitigation: Due to their powerful electronic components, wind turbines can introduce harmonics into the electrical system. Harmonics can cause voltage distortion, which can result in equipment failure and increased losses. To reduce harmonics and preserve the quality of the supplied power, active and passive filtering techniques, such as harmonic filters and power factor correction, are employed.
- Flicker control: Flicker is the rapid and perceptible variation in light intensity caused by voltage fluctuations. Variable-speed wind turbines can contribute to flickering due to fluctuations in power output. Advanced control strategies, including pitch control and reactive power compensation, are implemented to reduce flickering and ensure a stable power supply.
- Voltage and frequency ride-through capability: Wind turbines must be able to withstand voltage and frequency fluctuations in the grid. This feature ensures that wind turbines remain connected to the grid during transient events, such as grid faults and voltage drops. During these events, sophisticated control systems and grid support capabilities are utilized to facilitate grid integration and maintain power quality.
- Power quality monitoring and analysis: Continuous monitoring and analysis of power quality parameters are required to identify and mitigate power quality issues. Power quality parameters such as voltage, current, harmonics, and others are measured by sophisticated monitoring systems. The collected data is analyzed to identify anomalies, diagnose power quality issues, and optimize system performance.

Utilizing active power filters, voltage regulators, and reactive power compensation systems is an example of techniques to improve power quality.

In addition, standards and guidelines such as IEEE 519 and IEC 61400-21 specify and recommend power quality requirements for wind energy systems.

5.3.4.2 *Stability Theory*

Stability theory is an essential component of wind energy systems because it assures the reliability and safety of the power grid. Stability refers to a power system's capacity to maintain a balanced and stable state despite disturbances and alterations in operating conditions. Stability theory focuses on analyzing and improving the dynamic behavior of the system to guarantee grid stability in the context of wind energy (Chandra *et al.*, 2021).

Wind energy systems have three primary aspects of stability theory, namely:

A. Voltage stability
   Voltage stability refers to the maintenance of acceptable voltage levels within the power system. Variations in wind speed can result in fluctuating power output from wind turbines that are connected to the grid. These fluctuations have the potential to impact the voltage profile of the grid and cause voltage instability. Voltage stability analysis entails evaluating the effect of wind power fluctuations on the grid voltage and implementing control strategies to maintain acceptable voltage levels.
B. Frequency stability
   Maintaining a balance between power generation and demand requires frequency stability. Wind energy is variable, and abrupt changes in wind conditions can lead to power system imbalances and frequency deviations. Frequency stability analysis involves evaluating the wind turbines' response to varying wind speeds and implementing control mechanisms to regulate power output and support grid frequency control.
C. Dynamic stability
   Dynamic stability refers to the power system's ability to maintain stable and synchronized operation under transient conditions, such as defects or sudden load changes. Wind turbines must be able to withstand and recuperate from grid disturbances to preserve system stability. Analyzing the response of wind turbines to disturbances, assessing their impact on system stability, and employing control strategies to improve dynamic performance comprise dynamic stability analysis.

Several methods are employed to assure the stability of wind energy systems, namely:

- Control systems: Innovative control systems are essential for sustaining stability. These systems monitor grid conditions, wind turbine efficacy, and grid dynamics continuously. To maintain stability, they provide real-time feedback and adjust control parameters such as power output, reactive power compensation, and pitch control.
- Power system stabilizers: Stabilizers of the power system are devices that enhance the dynamic stability of the power system by injecting additional control signals. These stabilizers improve the reduction of power oscillations and contribute to the maintenance of stability during disturbances.
- Energy storage systems: Energy storage systems, such as batteries or flywheels, can be used to support stability by injecting or absorbing power rapidly during transient events. These systems aid in maintaining the frequency and voltage of the grid within permissible parameters.
- Grid codes and standards: Grid codes and standards specify the technical requirements and performance criteria for grid-connected wind energy systems. Compliance with these protocols guarantees that wind turbines contribute to system stability and grid dependability.

5.4 Advancements in Wind Energy Technologies

Wind energy technology advancements have substantially contributed to the expansion and effectiveness of wind power generation. These developments have focused on a variety of facets, including materials and manufacturing techniques, turbine designs, control systems and sensors, smart grid integration, and energy storage. Let us examine each of these facets in greater depth.

A. Innovative Materials and Production Methods
Wind turbines' efficacy and durability have been significantly improved by the development of advanced materials and manufacturing techniques. Utilizing lightweight and high-strength materials, such as carbon fiber composites, has resulted in longer and more efficient turbine blades. Turbine component production has become more precise and cost-effective due to

the implementation of advanced manufacturing techniques, such as 3D printing and automated manufacturing processes.

B. Innovative Designs for Wind Turbines
Researchers and engineers have consistently developed innovative wind turbine designs with the goals of maximizing energy capture, enhancing efficiency, and decreasing costs. This includes the development of larger turbines with higher hub heights to access winds that are stronger and more consistent at higher altitudes. In addition, concepts such as offshore floating wind turbines and vertical-axis wind turbines (VAWTs) have received increased attention as potential deployment solutions.

C. Enhancements to Control Systems and Sensors
To optimize the operation and performance of wind turbines, control systems and sensors have witnessed considerable development. Based on wind conditions, sophisticated control algorithms and real-time monitoring systems enable the precise adjustment of turbine settings such as blade pitch and rotor speed. Sensors provide valuable information on wind speed, direction, turbulence, and other environmental factors, allowing for more precise control and decision-making.

D. Smart Grid Integration and Energy Storage
Integration of wind power into smart grids is becoming increasingly important for grid stability and effective power management. Smart grid technologies allow for improved communication and coordination between wind farms, power utilities, and consumers, thereby facilitating the integration of wind energy into the existing power infrastructure. In addition, energy storage systems, such as batteries and pumped hydro storage, help mitigate the intermittent nature of wind power by storing excess energy during times of high generation and discharging it when the demand is high or the wind conditions are suboptimal.

These developments have significantly enhanced the efficacy, dependability, and cost-effectiveness of wind energy systems. They have also increased the competitiveness of wind energy in the global energy market.

5.5 Future Research Directions

Future research directions in wind energy seek to improve the efficiency, dependability, and sustainability of wind power production. These directions comprise several essential exploration and development areas.

Let us examine a few of the most important prospective research directions in wind energy, namely:

- Offshore wind energy: Due to the availability of stronger and more consistent winds in offshore locations, offshore wind energy has great potential for large-scale power generation. Future research will concentrate on addressing the technical obstacles associated with offshore wind farms, such as floating wind turbine technology, littoral conditions, marine operations, and transmission infrastructure. In addition, the research seeks to optimize the design and operation of offshore wind farms to maximize energy production while minimizing environmental impacts.
- Wind–wave interaction: The interaction between wind and ocean waves offers synergistic energy extraction opportunities. Future research will investigate the creation of hybrid energy systems that utilize both wind and wave energy. This includes the investigation of wave energy conversion technologies, the optimization of turbine designs to withstand severe marine conditions, and the exploration of the integration of wave energy converters with offshore wind farms.
- Advanced materials for wind turbines: Continuing research into advanced materials has the potential to enhance the efficacy and durability of wind turbines. This includes the investigation of novel materials with enhanced mechanical properties, including carbon nanotubes, graphene, and intelligent materials. To reduce the environmental impact of turbine manufacturing and decommissioning, research also concentrates on developing sustainable and recyclable materials.
- Integration of wind energy and energy storage: The integration of wind energy with energy storage systems is an active area of research. The intermittent character of wind energy can be mitigated by combining it with advanced energy storage technologies, such as large-scale batteries, hydrogen storage, or compressed air energy storage. The objective of the research is to optimize the integration of wind power with energy storage to enhance grid stability, enable flexible power dispatch, and support the penetration of renewable energy.

- Data analytics and machine learning: The use of data analytics and machine learning techniques could revolutionize wind energy systems. Future research will concentrate on developing advanced algorithms and models to optimize the performance of wind farms, more precisely predict wind conditions, and enhance fault detection and maintenance strategies. This includes the application of data-driven techniques for condition monitoring, turbine performance optimization, and advanced control systems.

5.6 Conclusion

Wind energy is essential to fulfilling the world's ever-increasing energy demands. It provides a pure and renewable energy source, thereby reducing greenhouse gas emissions and reliance on fossil fuels. Wind energy is significant due to its ability to contribute to a more sustainable and environmentally benign energy mix, thereby mitigating climate change and promoting the transition to a low-carbon future. This chapter has provided an extensive overview of wind energy theories based on technological advancements. It began by summarizing the various theories discussed throughout the chapter and emphasizing their importance to the advancement of wind energy. Aeroelasticity, wind turbine design, control systems, wind farm layout, turbine placement, wake effect, grid integration, power quality, stability, and technological advancements in wind energy have been thoroughly investigated. These theories have contributed to the development of wind energy systems that are efficient, dependable, and sustainable.

Wind energy contains immense promise and potential for the future. Ongoing technological and scientific advancements are anticipated to enhance the efficacy and efficiency of wind turbines, making them more affordable and accessible. Continued innovation in materials, turbine design, control systems, and grid integration will contribute to the maximization of energy capture, the reduction of operational costs, and the improvement of the overall reliability and productivity of wind energy systems. In addition, the integration of wind power with emergent technologies such as energy storage, smart grids, and advanced data analytics will improve the flexibility, stability, and grid integration capabilities of wind energy. This integration will improve the administration of variable wind resources, optimize power dispatch, and allow for the reliable and seamless integration of wind energy into the existing energy infrastructure.

References

1. Hu, Z., Liu, Y., Shi, Y., Xu, G., An Experimental Study on Rotor Aerodynamic Noise Control Based on Active Flap Control. *Aerospace*, 10, 121, 2023.
2. Alom, N., Saha, U.K., Evolution and Progress in the Development of Savonius Wind Turbine Rotor Blade Profiles and Shapes. *J. Sol. Energy Eng.*, 141, 2018.
3. Astolfi, D., Byrne, R., Castellani, F., Estimation of the Performance Aging of the Vestas V52 Wind Turbine through Comparative Test Case Analysis. *Energies*, 14, 915, 2021.
4. Azam, A., Ahmed, A., Wang, H., Wang, Y., Zhang, Z., Knowledge structure and research progress in wind power generation (WPG) from 2005 to 2020 using CiteSpace based scientometric analysis. *J. Clean. Prod.*, 295, 126496, 2021.
5. Bhatti, M.M., Marin, M., Zeeshan, A., Abdelsalam, S.I., Editorial: Recent Trends in Computational Fluid Dynamics. *Front. Phys.*, 8, 2020.
6. Caglayan, D.G., Ryberg, D.S., Heinrichs, H., Linßen, J., Stolten, D., Robinius, M., The techno-economic potential of offshore wind energy with optimized future turbine designs in Europe. *Appl. Energy*, 255, 113794, 2019.
7. Carrasco, J.M., Franquelo, L.G., Bialasiewicz, J.T., Galvan, E., PortilloGuisado, R.C., Prats, M.A.M., Leon, J.I., Moreno-Alfonso, N., Power-Electronic Systems for the Grid Integration of Renewable Energy Sources: A Survey. *IEEE Trans. Ind. Electron.*, 53, 1002, 2006.
8. Chandra, A., Singh, G.K., Pant, V., Protection of AC microgrid integrated with renewable energy sources – A research review and future trends. *Electr. Power Syst. Res.*, 193, 107036, 2021.
9. Chaudhuri, A., Datta, R., Kumar, M.P., Davim, J.P., Pramanik, S., Energy Conversion Strategies for Wind Energy System: Electrical, Mechanical and Material Aspects. *Materials (Basel).*, 15, 1232, 2022.
10. Chavero-Navarrete, E., Trejo-Perea, M., Jáuregui-Correa, J.-C., Carrillo-Serrano, R.-V., Rios-Moreno, J.-G., Pitch Angle Optimization by Intelligent Adjusting the Gains of a PI Controller for Small Wind Turbines in Areas with Drastic Wind Speed Changes. *Sustainability*, 11, 6670, 2019.
11. Chiulan, E.-A., Anton, A., The (r)evolution of wind energy systems in Romania: state-of-the-art, new trends and challenges. *IOP Conf. Ser. Earth Environ. Sci.* 664, 12016, 2021.
12. Eltayesh, A., Castellani, F., Burlando, M., Bassily Hanna, M., Huzayyin, A.S., El-Batsh, H.M., Becchetti, M., Experimental and numerical investigation of the effect of blade number on the aerodynamic performance of a small-scale horizontal axis wind turbine. *Alexandria Eng. J.*, 60, 3931, 2021.
13. Famoso, F., Brusca, S., D'Urso, D., Galvagno, A., Chiacchio, F., A novel hybrid model for the estimation of energy conversion in a wind farm combining wake effects and stochastic dependability. *Appl. Energy*, 280, 115967, 2020.

14. Guoqiang, L., Weiguo, Z., Yubiao, J., Pengyu, Y., Experimental investigation of dynamic stall flow control for wind turbine airfoils using a plasma actuator. *Energy*, 185, 90, 2019.
15. Hannan, M.A., Al-Shetwi, A.Q., Mollik, M.S., Ker, P.J., Mannan, M., Mansor, M., Al-Masri, H.M.K., Mahlia, T.M.I., Wind Energy Conversions, Controls, and Applications: A Review for Sustainable Technologies and Directions. *Sustainability*, 15, 3986, 2023.
16. Hart, E., Clarke, B., Nicholas, G., Kazemi Amiri, A., Stirling, J., Carroll, J., Dwyer-Joyce, R., McDonald, A., Long, H., A review of wind turbine main bearings: design, operation, modelling, damage mechanisms and fault detection. *Wind Energy Sci.*, 5, 105, 2020.
17. Alhajj Hassan, F., Mahmoud, M., Almohammed, O.A.M., Analysis of the Generated Output Energy by Different Types of Wind Turbines. *J. Human, Earth, Futur.*, 1, 181, 2020.
18. Hou, P., Zhu, J., MA, K., YANG, G., HU, W., CHEN, Z., A review of offshore wind farm layout optimization and electrical system design methods. *J. Mod. Power Syst. Clean Energy*, 7, 975, 2019.
19. Campbell, L.K., City of Forests, City of Farms, http://dx.doi.org/10.7591/cornell/9781501707506.001.0001, (2018).
20. Syahputra, R., Soesanti, I., Power System Stabilizer model based on Fuzzy-PSO for improving power system stability, http://dx.doi.org/10.1109/icamimia.2015.7508015, (2015).
21. Jiang, C., Moreau, D., Fischer, J., Doolan, C., Additively Manufactured Sound-Absorbing Porous Structures for Airfoil Trailing-Edge Noise Control. *J. Aerosp. Eng.*, 34, 2021.
22. Kharrazi, A., Sreeram, V., Mishra, Y., Assessment techniques of the impact of grid-tied rooftop photovoltaic generation on the power quality of low voltage distribution network - A review. *Renew. Sustain. Energy Rev.*, 120, 109643, 2020.
23. Kishor, S., Rajesh, K., Rajendrn, S., Ramkumar, A., Arunkumar, T., A Comparative Analysis of Maximum Power Point Tracking Algorithms Applied to Hybrid Wind and Solar System, http://dx.doi.org/10.1109/icac3n53548.2021.9725451, (2021).
24. Kumar, G.B.A., Shivashankar, Optimal power point tracking of solar and wind energy in a hybrid wind solar energy system. *Int. J. Energy Environ. Eng.*, 13, 77, 2021.
25. Kumar, R., Ismail, M., Zhao, W., Noori, M., Yadav, A.R., Chen, S., Singh, V., Altabey, W.A., Silik, A.I.H., Kumar, G., Kumar, J., Balodi, A., Damage detection of wind turbine system based on signal processing approach: a critical review. *Clean Technol. Environ. Policy*, 23, 561, 2021.
26. Kwok, K.C.S., Hu, G., Wind energy system for buildings in an urban environment. *J. Wind Eng. Ind. Aerodyn.*, 234, 105349, 2023.
27. Lai, C.-M., Teh, J., Network topology optimisation based on dynamic thermal rating and battery storage systems for improved wind penetration and reliability. *Appl. Energy*, 305, 117837, 2022.

28. Lerch, M., De-Prada-Gil, M., Molins, C., Collection Grid Optimization of a Floating Offshore Wind Farm Using Particle Swarm Theory. *J. Phys. Conf. Ser.*, 1356, 12012, 2019.
29. Liu, J., Hu, H., Yu, S.S., Trinh, H., Virtual Power Plant with Renewable Energy Sources and Energy Storage Systems for Sustainable Power Grid-Formation, Control Techniques and Demand Response. *Energies*, 16, 3705, 2023.
30. Maza, M.S., Preidikman, S., Flores, F.G., A cost-effective algorithm for the co-simulation of unsteady and non-linear aeroelastic phenomena. *J. Fluids Struct.*, 118, 103838, 2023.
31. McKenna, R., Pfenninger, S., Heinrichs, H., Schmidt, J., Staffell, I., Bauer, C., Gruber, K., Hahmann, A.N., Jansen, M., Klingler, M., Landwehr, N., Larsén, X.G., Lilliestam, J., Pickering, B., Robinius, M., Tröndle, T., Turkovska, O., Wehrle, S., Weinand, J.M., Wohland, J., High-resolution large-scale onshore wind energy assessments: A review of potential definitions, methodologies and future research needs. *Renew. Energy*, 182, 659, 2022.
32. Menezes, E.J.N., Araújo, A.M., Wind turbine structural control using H-infinity methods. *Eng. Struct.*, 286, 116095, 2023.
33. Mu, Z., Tong, G., Xiao, Z., Deng, Q., Feng, F., Li, Y., Arne, G., Study on Aerodynamic Characteristics of a Savonius Wind Turbine with a Modified Blade. *Energies*, 15, 6661, 2022.
34. Nazir, M.S., Ali, N., Bilal, M., Iqbal, H.M.N., Potential environmental impacts of wind energy development: A global perspective. *Curr. Opin. Environ. Sci. & Heal.*, 13, 85, 2020.
35. Nishino, T., Dunstan, T.D., Two-scale momentum theory for time-dependent modelling of large wind farms. *J. Fluid Mech.*, 894, 2020.
36. O'Neill, S., Giant Turbines Poised to Claim Offshore Wind. *Engineering*, 7, 894, 2021.
37. Sayahi, K., Kadri, A., Bacha, F., Marzougui, H., Implementation of a D-STATCOM control strategy based on direct power control method for grid connected wind turbine. *Int. J. Electr. Power & Energy Syst.*, 121, 106105, 2020.
38. Shah, S.A.A., Solangi, Y.A., A sustainable solution for electricity crisis in Pakistan: opportunities, barriers, and policy implications for 100% renewable energy. *Environ. Sci. Pollut. Res.*, 26, 29687, 2019.
39. Sitharthan, R., Sundarabalan, C.K., Devabalaji, K.R., Yuvaraj, T., Mohamed Imran, A., Automated power management strategy for wind power generation system using pitch angle controller. *Meas. Control.*, 52, 169, 2019.
40. Subbulakshmi, A., Verma, M., Keerthana, M., Sasmal, S., Harikrishna, P., Kapuria, S., Recent advances in experimental and numerical methods for dynamic analysis of floating offshore wind turbines — An integrated review. *Renew. Sustain. Energy Rev.*, 164, 112525, 2022.
41. Sun, H., Qiu, C., Lu, L., Gao, X., Chen, J., Yang, H., Wind turbine power modelling and optimization using artificial neural network with wind field experimental data. *Appl. Energy*, 280, 115880, 2020.

42. Torregrosa, A.J., García-Cuevas, L.M., Quintero, P., Cremades, A., On the application of artificial neural network for the development of a nonlinear aeroelastic model. *Aerosp. Sci. Technol.*, 115, 106845, 2021.
43. Vakacharla, V.R., Gnana, K., Xuewei, P., Narasimaharaju, B.L., Bhukya, M., Banerjee, A., Sharma, R., Rathore, A.K., State-of-the-art power electronics systems for solar-to-grid integration. *Sol. Energy*, 210, 128, 2020.
44. Vali, M., Petrović, V., Boersma, S., van Wingerden, J.-W., Pao, L.Y., Kühn, M., Adjoint-based model predictive control for optimal energy extraction in waked wind farms. *Control Eng. Pract.*, 84, 48, 2019.
45. Wilberforce, T., Olabi, A.G., Sayed, E.T., Alalmi, A.H., Abdelkareem, M.A., Wind turbine concepts for domestic wind power generation at low wind quality sites. *J. Clean. Prod.*, 394, 136137, 2023.
46. Xie, F., Aly, A.-M., Structural control and vibration issues in wind turbines: A review. *Eng. Struct.*, 210, 110087, 2020.
47. Xu, B., Luo, X., Egusquiza, M., Ye, W., Liu, J., Egusquiza, E., Chen, D., Guo, P., Nonlinear modal interaction analysis and vibration characteristics of a francis hydro-turbine generator unit. *Renew. Energy*, 168, 854, 2021.
48. Zhao, J., Pei, J., Yuan, J., Wang, W., Energy-saving oriented optimization design of the impeller and volute of a multi-stage double-suction centrifugal pump using artificial neural network. *Eng. Appl. Comput. Fluid Mech.*, 16, 1974, 2022.
49. Zhao, X., Wang, C., Su, J., Wang, J., Research and application based on the swarm intelligence algorithm and artificial intelligence for wind farm decision system. *Renew. Energy*, 134, 681, 2019.

6

Wind Energy Hybrid Power Generation System with Hydrogen Storage

Mehmet Bugdayci[1,2]* and Mesut Yilmazoglu[1]

[1]*Chemical Engineering Department, Faculty of Engineering, Yalova University, Yalova, Turkey*

[2]*Construction Technology Department, Vocational School, Istanbul Medipol University, Istanbul, Turkey*

Abstract

The focus of this study is to determine the optimum use of wind energy and to find a complex system conditions in which hydrogen storage is possible from this power source. In this chapter, the superior properties of hydrogen in energy transport and various hydrogen production technologies are emphasized. It then explores the potential of combining wind power with hydrogen storage to create a more efficient and weather-independent complex power production process. The proposed system is designed to maximize energy production and minimize fluctuations in energy output by storing excess energy generated at higher wind speeds for later use. The chapter also discusses the key components of the recommended process, including the generator, electrolysis device, H storage tank, reproductive cell, power control system, and wind turbine. Finally, the chapter presents a theoretical study on the proposed system and discusses its potential applications in providing a more sustainable and eco-friendly source of electricity for areas with limited access to electricity or clean energy. Overall, the chapter provides valuable insights into the development and institutionalization of a hydrogen storage air energy complex power production process as a promising solution to the challenges of relying solely on wind power as a power source.

Keywords: Green resources, hydrogen storage, wind energy, hybrid power generation, hydrogen storage–wind energy

**Corresponding author*: mehmet.bugdayci@yalova.edu.tr

Inamuddin, Tariq Altalhi and Mohammad Luqman (eds.) Wind Energy Storage and Conversion: From Basics to Utilities, (115–132) © 2024 Scrivener Publishing LLC

6.1 Introduction

Interest in eternal energy providers is increasing day by day as society continues to shift toward more sustainable and environmentally friendly energy options. While these sources offer many benefits, such as reduced emissions and decreased dependence on non-renewable resources, they do come with certain challenges. The primary challenges associated with renewable sources is their intermittency. Unlike fossil fuels, the energy provided from natural sources (sun, water, and wind) is affected by factors such as time and weather conditions. To overcome this issue, energy storage is necessary to ensure that the energy produced during low demand periods can be stored and utilized during high demand periods. It is possible to achieve this with hydroelectric pump, battery, and some energy storage devices.

By integrating natural energy sources with energy storage technology, a dependable and steady energy system can be established, rendering it ideal for implementation in remote and isolated settings. In instances such as remote regions where conventional electricity generation is impractical, establishing an independent power plant reliant on fossil fuels would prove costly and environmentally detrimental. In contrast, the amalgamation of renewable energy and energy storage offers a more economical and eco-friendly alternative. In periods when the energy demand is low, it is essential to store energy and to use it without any problems in increasing demand. While green sources are intermittent, the utilization of energy storage technologies can help overcome this challenge and make them a more reliable and stable energy source. This not only benefits the environment but also provides a cost-effective and sustainable solution for remote and stand-alone applications [1–3].

Considering the environmental and sustainable energy trend, the economic benefits of green resources are of great importance. While diesel generators have traditionally been used in stand-alone systems, their costs can be high, both in terms of installation and ongoing maintenance. In contrast, renewable energy systems are capable of producing low-cost energy, especially when combined with advanced energy storage devices. Compared to diesel engines, the costs of renewable energy generators are very low. This is because green sources have no fuel costs, and their maintenance costs are relatively low [4]. Additionally, the costs of energy storage technologies have been decreasing in recent years, making them increasingly affordable. In rural areas where there is no electricity infrastructure, storing energy from green sources will result in a much more advantageous cost. In addition, green resources will lead to lowering the carbon footprint and promoting sustainable energy sources. The advantages arising from costs and environmental factors make green energy storage and use systems more usable both in the countryside and in cities. Although wind energy is a strong renewable energy source, its variability can pose challenges [5]. Wind speeds can fluctuate greatly from moment to moment and season to season, making energy storage essential if wind power is to meet significant energy demands. Historically, energy storage systems have been inefficient and costly, preventing wind power from becoming a practical energy source. However, advances in technology have made wind-to-electricity conversion more attainable. Experiments with wind farms have shown that the most feasible approach to using electrical energy generated by wind power is to supplement it with other forms of energy in a hybrid system. By mixing wind power with other green sources, such as sun or hydro, the fluctuations in wind energy can be offset, resulting in a more reliable and efficient energy supply [6]. Overall, while wind energy's variability has historically posed challenges, advancements in technology and the use of hybrid energy systems offer energy amount of air systems to become a practical and efficient energy source [4–7].

The declining reserves of conventional fuels, lack of water reserves, unpredictable prices, high costs of diesel generators, and the threat of global warming have raised concerns about the utilization of conventional power sources. In developing countries, nearly 1.5 billion people in the world have difficulty in accessing electricity, leading to increased deforestation as they rely on wood for energy. Alternative energy (AE) sources such as air, sun, water, biomass, fuel cells, and tidal power can address these concerns as they are renewable, non-polluting, and readily available. Figure 6.1 shows a list of the features of AE and conventional energy sources. AE sources offer benefits such as reduced carbon emissions, improved power reliability, and lower operational and maintenance costs. Since the mid-1970s, there have been studies on the components of the AE process in the literature. To improve living standards in electricity-deprived remote areas, many different energy process designs consisting of green energy systems have been revealed [8–11].

Realization of processes containing AE in isolation to meet a load demand can be costly and result in oversized systems. To address this, system designers are exploring the option of integrated operation, which involves combining two or more AE sources with energy storage to create hybrid systems that can generate power consistently. Figure 6.2 provides various methods for storing energy. This section discusses the construction and operation of these hybrid systems and the software tools used for

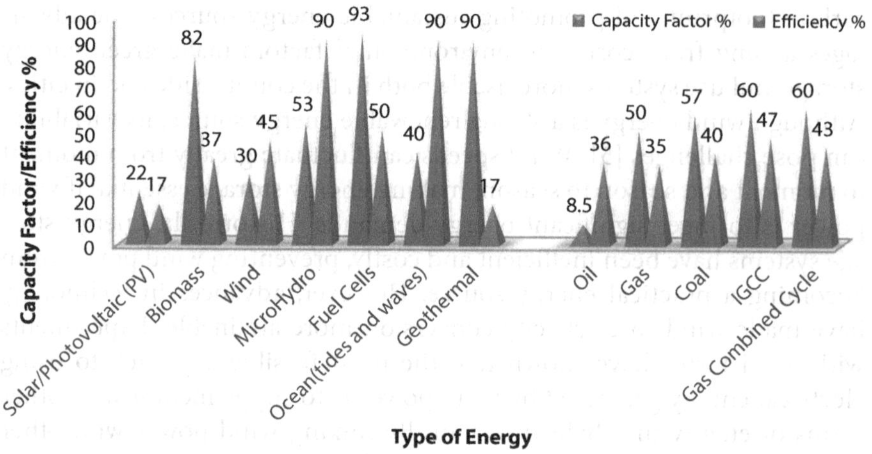

Figure 6.1 Capacity and efficiency of various energy types.

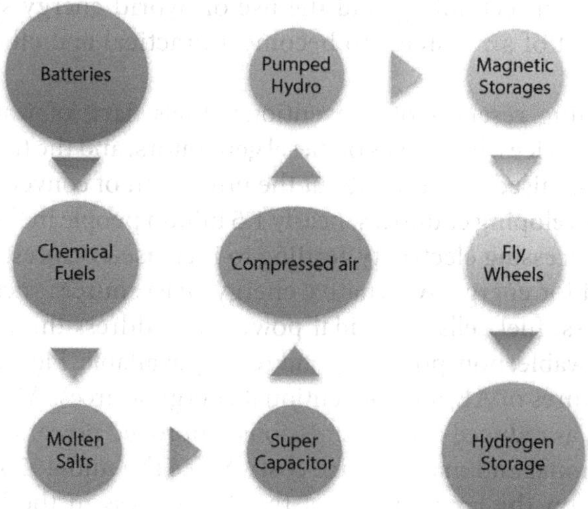

Figure 6.2 Methods for energy storage.

their analysis and provides examples of different hybrid systems from literature. The section also covers the potential and implementation of renewable energy sources worldwide, including solar, micro hydro, fuel cells, and biomass [10, 12–14].

Fossil fuels come to the forefront with their easy storage and transportation and high energy holding properties. However, the conversion and transportation of green resources from water and biowaste into electricity is problematic. Electricity is an energy form that is easy to use and transmit, so its use is high. Energy storage can help with this problem in several ways, such as regulating system frequency, providing spinning reserve, peak shaving, load leveling, and storing surplus electricity generated by renewables. In the process of storing energy, advantages are obtained with quality, minimization of losses in transport, and economic savings. Even in regions with developed and complex lines, storage in rural areas is only provided by hydroelectric power plants. Increasing the storage capacity of batteries is a feature expected and desired by every stakeholder. The inclusion of green energy in the system will further increase the demand for storage. It will be sufficient to use smaller batteries compared to the aforementioned processes to store the energy obtained from the air [15–17].

The storage of green energy is perhaps the most important factor for this process. Since the energy provided by air and sun varies according to climate and time conditions, it is important to provide optimum conditions in terms of efficiency. The diagnosis of optimum storage conditions will ensure a smooth process by securing excess energy and releasing it during periods of high demand. There are many benefits of storing energy. Enhanced potential and reliability can result from more stable and predictable power output. Reduced transmission and power losses can result from the ability to store energy closer to the point of use. Finally, tactical benefits can be gained through greater siting and fuel flexibility [18–20].

The energy transmission lines of the countries bordering the English Channel have been designed in coordination with each other. More costly and complex facilities are needed to meet the high demands occurring here. In these regions, the system is supported by hydroelectric power plants and simple battery facilities. In this context, the improvement in storage capacity will be welcomed in many developed countries.

With the demand from green energy, there is an increasing demand for energy storage. The capacity of these energy sources will vary according to the amount of wind blend and the condition of the solar panels. The energy to be stored from these sources will, of course, be lower than with conventional sources. Accordingly, by making correct deployments, buildings, shopping malls or workplaces will be fed easily from green sources [21, 22].

When viewed from a broader spectrum, energy storage has a critical role in the use of environmentalists and green energies. While economic viability remains a challenge, the potential benefits of energy storage are significant, and continued research and development will be critical to unlocking its full potential.

6.2 Hydrogen Storage Systems

After the industrial revolution, the need for energy has increased a lot, and as a result, reducing carbon emissions has emerged as a necessity. At this stage, hydrogen stands out as a clean energy carrier. Various processes have been developed for hydrogen production; electrolysis, photocatalysts, thermolysis, and bio-fermentation. In the study of Dawood *et al.*, hydrogen has an incredible preference over fossil sources due to its gravimetric energy density and high caloric value [23–27]. This is illustrated in Figure 6.3.

Hydrogen stands out as the best alternative to traditional energy sources, with its high heating value of 120 MJ/kg and its nature-friendly structure. In fact, hydrogen is a superior alternative to other energy sources, as demonstrated in Figure 6.4.

Although hydrogen is one of the three most abundant elements on earth, it occurs not in elemental form but in compound form. Therefore, various processes are needed to bring hydrogen from the compound form to the elemental form. The H production process is performed through different methods, and then it needs to be stored and transported to its destination. However, storing hydrogen is challenging due to its low density and high flammability, which make conventional compressed storage risky [28, 29].

When focusing on sustainable energy sources, hydrogen is one of the prominent materials. The storage processes of this material should have standardization. In this review, solid-state energy storage systems by absorption are examined and the process details are discussed. The study explores the gradual progression of material properties that can be achieved through customization of alloy, providing adjustments in material PCT (pressure–composition–temperature) properties to meet

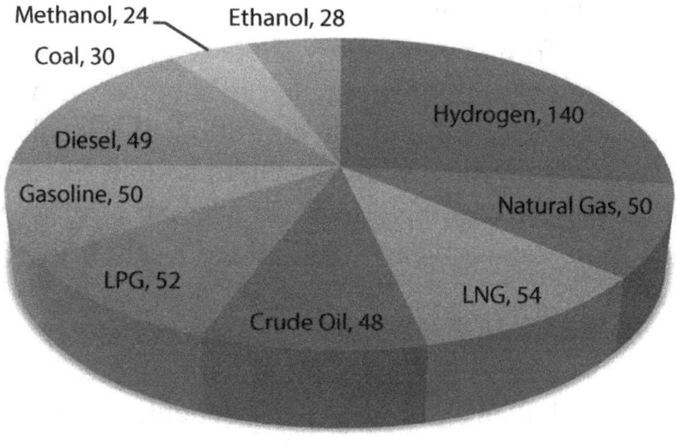

Figure 6.3 Available sources' energy values (MJ/kg).

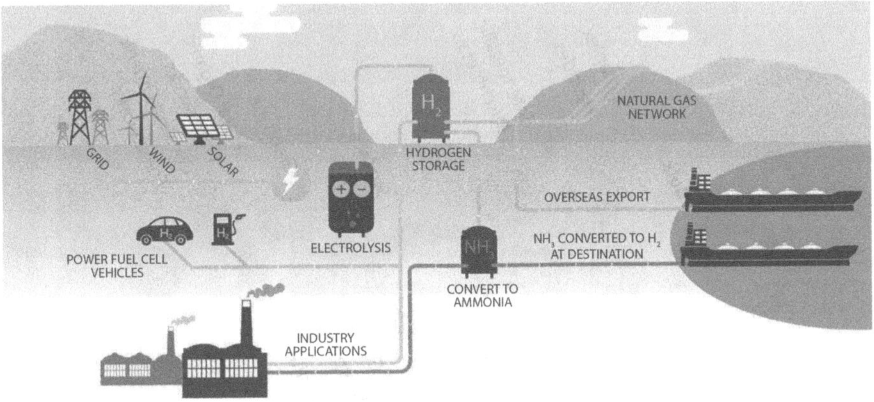

Figure 6.4 Hydrogen production and transportation.

specific applications. Alloys with cubic volume-centered crystal lattice and complex hydrides are among the stakeholders that make up this material. The main purpose is to bring the storage systems created by the adsorption process to the literature with the details [30–34].

6.2.1 Solid-State Hydrogen Storage in Materials

These cubic volume-centered materials appear as H storage materials. Alloys can store hydrogen as compressed by adding it into the structure with the absorption technique. These technologies are important processes for both the storage of hydrogen and its high-efficiency transmission. In the research and development activities carried out on the subject, it is aimed to develop the optimum cycle values of hydrogen, fast absorption ability, low pressure, and system temperature. The storage and delivery of hydrogen with a safe process has become a topic of great interest in the last decade. The issues that energy transport focuses on are vehicles using H based on green energy sources, industries, batteries, and homes [35–37]. There are two main categories of hydrogen storage systems based on absorption.

In H storage systems where solid-state absorption is made, +1 valence atoms form metal hydrides to neutralize the bond energy and are absorbed by a cubic volume-centered lattice. In the absorption process, which proceeds depending on the surface diffusion, H atoms settle on the gaps and cracks and accumulate on the surface. The absorption-based H storage systems shown in Figure 6.5 are divided into two as surface and solid-state absorption. Accordingly, this research focuses on the absorption characteristics of materials with H [30, 38–41].

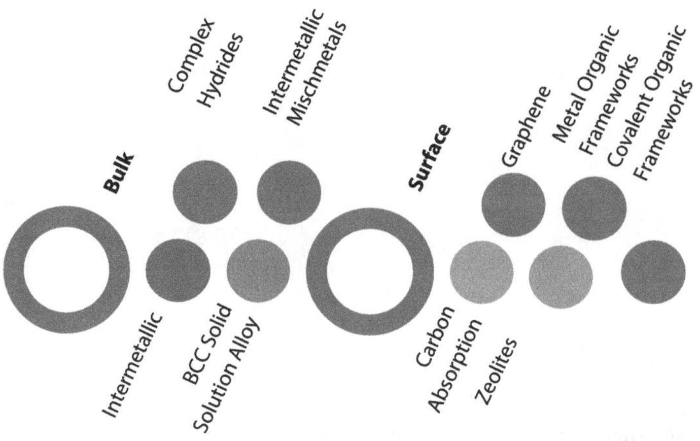

Figure 6.5 Hydrogen absorption storage techniques.

In the production of metal hydrides, it is essential to obtain intermetallics in accordance with the laws of thermodynamics. The first of the elements combined here consists of weak hydriding, and the second consists of the element with strong properties. These alloys have attracted great attention in recent years with their superior properties in hydrogen storage. The formation reactions of these metal hydrides show an ando- or exothermic character according to their Gibbs free energies. Accordingly, as a result of the reaction, a decrease or increase in the temperature of the environment may occur. It provides an environment suitable for use in environments such as pumps, transformers, and compressors, where this diathermal environment is important [42].

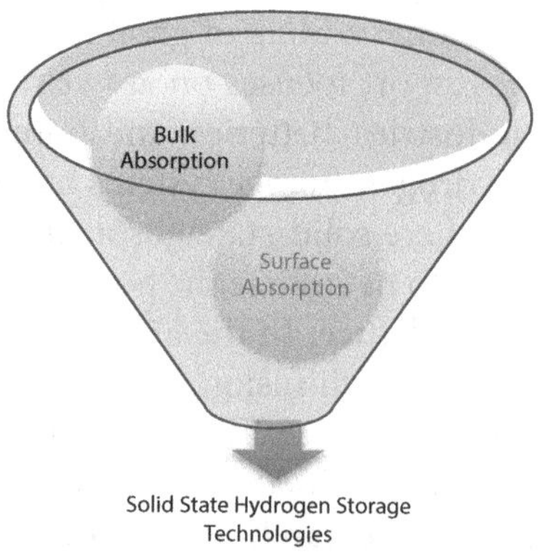

Figure 6.6 Solid state hydrogen storage technology (Author's own).

6.3 Wind Energy Systems

Green energy sources contribute to both economic and social developments by reducing the need for fossil fuels. High pressure and low pressure centers caused by global warming have caused a significant increase in the amount of wind. This encourages the use of wind energy. This energy has become a good energy source alternative as an endless source against zero carbon footprint limited fossil fuels. The demand for electricity continues with a rising trend since the industrial revolution. It is getting more and more difficult to meet this demand with known methods. While fossil fuels meet the basic energy needs in the past century, the harmful effects of these fuels on the environment cannot be denied. Therefore, it is important to turn to environmentally friendly green energy sources. These resources include solar, tidal, wave, geothermal, and wind energy. Among these, the wind obtained from the air stands out with its natural structure and affordable cost. Figure 6.7 shows the wind harvests of the leading centers in the world.

Wind turbines are designed to withstand extreme wind conditions, but they are not able to endure high rotational speeds or torques. These turbines are built with a cut-out speed, beyond which the brakes will bring the turbine to a halt. Without a cut-out speed, the turbine blades and other components are subjected to immense forces due to increased aerodynamic torques and rotational speeds, leading to damage. Wind turbines come in different types and can be classified into two groups based

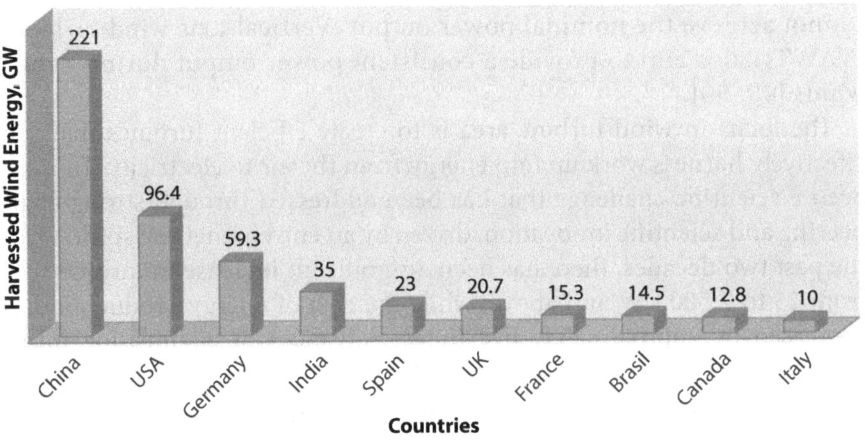

Figure 6.7 Wind energy harvest by countries.

on their axis of rotation: horizontal-axis wind turbines (HAWTs) and vertical-axis wind turbines (VAWTs). HAWTs are more widely used and have a higher expected mass compared to the lesser-known VAWT technologies [43–45].

Lately, there has been a rising curiosity in performing a comparative analysis between horizontal-axis wind turbines (HAWTs) and vertical-axis wind turbines (VAWTs). Although HAWTs have been the main subject of investigation when evaluating the existing state of wind power, there is a growing trend now to examine both VAWTs and HAWTs side by side. The speed of the wind is an important parameter in the application processes of wind turbines, both in obtaining the harvest and with the possibility of causing damage to the system. Appropriate control procedures need to be developed according to environmental conditions. These procedures are basically divided into two as pitch-regulated and fixed-speed turbines. Pitch systems are activated when there is a strong wind, while fixed-speed turbines change the angle to bring the rotation speed to the appropriate point to provide efficient torque production [46, 47].

Pitch-controlled wind turbines are processed to reduce power production above a certain wind speed by decreasing their rotational speed and aerodynamic torques when the blades start to stall. The stall regulation requires less capital cost and has fewer maintenance and moving parts compared to yaw regulation. Both pitch-controlled and stall-controlled wind turbines have on-site systems to stop the turbine at high wind speeds. However, pitch-controlled systems use active pitch control for the blades, enabling them to maintain a consistent power output beyond the rated wind speed, while stall-controlled systems cannot achieve the nominal power output. Vertical-axis wind turbines (VAWTs) also aim to provide a consistent power output during strong winds [48–50].

The focus in wind turbine area is to create efficient turbines that can effectively harness work up into energy from the air to electricity. This has been a scientific challenge that has been addressed through strong engineering and scientific innovation, driven by an entrepreneurial spirit. Over the past two decades, there has been a significant increase in turbine size, from 25 to 2,500 kW and above, while the cost of energy production has decreased by approximately five times. Advances in engineering infrastructure and computational devices have allowed for the correlation of machine size and volume to be improved [51–53].

6.4 Wind Energy Hybrid Power Generation System with Hydrogen Storage

Among the green energy sources, wind stands out among the sustainable energies that have become a trend with its convenience in application and its affordable cost compared to other complex systems [54, 55]. However, it should be noted that wind power generation cannot be completely continuous and can be affected by a variety of wind speed and direction. This complicates the singular usefulness of wind power systems as a power source [56–58].

To address the problems of these individual wind systems, hybrid power generation systems have been developed, i.e., the integration of wind power with diverse renewable energy sources, including sun power, results in a more dependable and continuous power supply. Power generation processes consisting of a minimum of two components are designed for energy production with maximum efficiency and to minimize fluctuations in energy output that can occur when relying on a single energy source [59–61].

One of the most effective hybrid power generation systems among these hybrid systems is the hydrogen storage hybrid power generation system. This system uses wind energy to generate electricity, which is then stored as hydrogen gas [62]. The use of $H_{2(g)}$ can have a positive effect in electricity generation to make the system efficient and to save it from adverse weather conditions as much as possible [62, 63].

The efficient hydrogen storage wind energy minimum two-component power harvesting method consists of several primary components. The most important component is the wind turbine, which converts wind energy directly into mechanical energy [64]. The converted mechanical energy is then converted into electrical energy by a generator. With the electrolysis method, electricity is produced by separating water into two components, H_2 and O. Hydrogen (g) is then stored in tanks until it is needed to generate electricity. When electricity is required, hydrogen (g) is run through a fuel cell that combines hydrogen with oxygen to produce electricity and water [65].

H store hybrid power generation systems provide advantages in many aspects. The system provides a reliable and efficient power supply even when wind speeds are low or the weather and additional parameter conditions are unfavorable. In addition, the system can also store excess energy generated at higher wind speeds, which can then be used at lower wind speeds.

This helps stabilize the energy output of the system and provides much more energy source [66].

In addition, the hydrogen storage–wind power hybrid power generation system is environmentally friendly, producing no emissions or pollutants [67]. It also helps reduce dependency on fossil fuels and contributes to a more sustainable energy future. As a result, when wind energy is considered as a source in hybrid green energy systems, the solution of the problems of this issue is promising for the future. While providing a reliable and consistent power supply, it is also environmentally friendly and contributes to a more sustainable energy future.

6.4.1 Design and Optimization of a Wind Energy Hybrid Power Generation System with Hydrogen Storage

The proposed hydrogen storage–wind energy two-component power system consists of several key components, including a generator, electrolyzes, hydrogen storage tank, fuel cell, power control system, and wind turbine [68].

Yang and Aydin presented a two-component power system that uses air-based energy and H storage in their theoretical study, which was one of the first of its kind in the literature. Initially, they calculated the available wind energy and evaluated the yield of the air-based energy conversion system. After revising their windmill model, they assessed the wind power density and electric power output. In cases where the supply is more than the demand, the excess electrical power is fed into the electrolysis cell and continues to be used in the system again. The $H_{(g)}$ released during this process can be stored as liquid or condensed gas. The gas produced here can be stored in pressure vessels, used as fuel, in steam production, or in power generation turbines [69]. A large proportion of the world's population does not have access to electricity or clean drinking water. Off-grid hybrid renewable energy systems, particularly those using wind and solar power, are being considered as a solution to address both of these needs. However, designing and implementing these systems is complex, and achieving an optimal balance between wind and solar resources and proper storage requires careful engineering solutions. Maliki *et al.* conducted a study where six hybrid systems were designed and modeled to increase the availability of freshwater and meet the energy demands of the population. An improved bees algorithm was used to optimize the design of these systems and compared with the fit search algorithm. The results showed that GHRES-based battery energy storage was more cost-effective than GHRES-based hydrogen energy storage. The hybridization of solar, battery, and ROD was the most cost-effective energy system with minimal power supply loss. Overall, the proposed method showed promising results in achieving efficient off-grid hybrid renewable energy systems [70].

Renewable energy sources like wind and solar are intermittent, which means that they are not consistently available. This unpredictability can lead to mismatches in the power grid. However, the diversity in the H production process can overcome these problems. $H_{(g)}$ can be used as a raw material with the ability to feed back into electricity by electrolysis. It can also be used for heating buildings and as fuel for vehicles. Therefore, hydrogen production and storage can help address the challenges of intermittency in renewable energy sources [71–73].

In the green energy synthesis process, many studies are carried out to determine the optimum conditions for its use in the $H_{(g)}$ phase. The study used Markov decision process theory to determine the best policies for day-to-day decisions such as determining how much energy to store as hydrogen, how much to buy and sell on the electricity market, and how much hydrogen to sell for use as gas. In the feasibility study, electricity transmission conditions, cost, and H market and H purchase conditions were examined. In the environment identified in northern Europe, the H valley was established, and the necessary R&D studies were carried out. The results showed that earnings could be increased by 51%. For a 4.5-MW wind turbine, this translates to an increase in revenues of €126,000 per year. The study also emphasized that H purchase deals will be critical in ensuring a successful energy transportation [74].

In Fan's study, they developed a hijritic system consisting of fossil fuel, air, and PV in order to obtain energy from the wind. The goal of the system was to support the sustainable development of the region while also addressing the issue of air and sun power outages in Hami [75]. In the study of Fan *et al.*, air and PV were added to the fossil fuel industry. Energy production is provided by taking power from fossil fuel (coal), wind, and PV. Additionally, wind and PV power are used to generate hydrogen, which helps to mitigate the negative impact of the energy sources' stochastic, intermittent, and volatile properties on the grid. The presence of wind energy in the system eliminates the need for complex air cleaning units. While the amount of CO_2 produced by the coal-derived Boduard reaction decreases significantly with green sources, the reaction products are evaluated in methanol synthesis [75].

6.5 Conclusion

In conclusion, the wind energy hybrid power generation system with hydrogen storage is a promising solution to the challenges of relying solely on wind power as a power source. This system provides a reliable and efficient power supply even when wind speeds are low or weather conditions are unfavorable. The system can also store excess energy generated at higher wind speeds, which can then be used at lower wind speeds, stabilizing the energy output of the system. In addition, the hydrogen storage-wind power hybrid power generation system is environmentally friendly, producing no emissions or pollutants and contributing to a more sustainable energy future. With further development and optimization, this technology has the potential to play an important role in meeting the growing global demand for clean and renewable energy.

References

1. Shakya, B.D., Aye, L., Musgrave, P., Technical feasibility and financial analysis of hybrid wind-photovoltaic system with hydrogen storage for Cooma. *Int. J. Hydrogen Energy*, 30, 9, 2005.
2. Clark, W., Isherwood, W., Distributed generation: Remote power systems with advanced storage technologies. *Energy Policy*, 32, 1573, 2004.
3. Dell, R.M., Rand, D.A.J., Energy storage - A key technology for global energy sustainability. *J. Power Sources*, 100, 2, 2001.
4. Schrotenboer, A.H., Veenstra, A.A.T., uit het Broek, M.A.J., Ursavas, E., A Green Hydrogen Energy System: Optimal control strategies for integrated hydrogen storage and power generation with wind energy. *Renew. Sustain. Energy Rev.*, 168, 112744, 2022.
5. Yang, W.J., Aydin, O., Wind energy-hydrogen storage hybrid power generation. *Int. J. Energy Res.*, 25, 449, 2001.
6. Raju, M., Khaitan, S.K., System simulation of compressed hydrogen storage based residential wind hybrid power systems. *J. Power Sources*, 210, 303, 2012.
7. Zhou, T., Francois, B., Modeling and control design of hydrogen production process for an active hydrogen/wind hybrid power system. *Int. J. Hydrogen Energy*, 34, 21, 2009.
8. Kornelakis, A., Multiobjective Particle Swarm Optimization for the optimal design of photovoltaic grid-connected systems. *Sol. Energy*, 84, 2022, 2010.
9. Hepbasli, A., Low exergy (LowEx) heating and cooling systems for sustainable buildings and societies. *Renew. Sustain. Energy Rev.*, 16, 73, 2012.
10. Aziz, M.S., Ahmed, S., Saleem, U., Mufti, G.M., Wind-hybrid power generation systems using renewable energy sources-A review. *Int. J. Renew. Energy Res.*, 7, 111, 2017.
11. Sharma, A., Saxena, A., Sethi, M., Shree, V., Varun, Life cycle assessment of buildings: A review. *Renew. Sustain. Energy Rev.*, 15, 871, 2011.

12. Dujardin, J., Kahl, A., Kruyt, B., Bartlett, S., Lehning, M., Interplay between photovoltaic, wind energy and storage hydropower in a fully renewable Switzerland. *Energy*, 135, 513, 2017.
13. Kamel, R.M., Standalone micro grid power quality improvement using inertia and power reserves of the wind generation systems. *Renew. Energy*, 97, 572, 2016.
14. Yang, Y., Lian, C., Ma, C., Zhang, Y., Research on energy storage optimization for large-scale PV power stations under given long-distance delivery mode. *Energies*, 13, 27, 2019.
15. Liu, L. qun, Wang, Z. xin, Zhang, H. qiang, Xue, Y. cheng, Solar energy development in China-A review. *Renew. Sustain. Energy Rev.*, 14, 301, 2010.
16. Jamal, T., Urmee, T., Calais, M., Shafiullah, G.M., Carter, C., Technical challenges of PV deployment into remote Australian electricity networks: A review. *Renew. Sustain. Energy Rev.*, 77, 1309, 2017.
17. Zhang, Y., Ma, C., Lian, J., Pang, X., Qiao, Y., Chaima, E., Optimal photovoltaic capacity of large-scale hydro-photovoltaic complementary systems considering electricity delivery demand and reservoir characteristics. *Energy Convers. Manag.*, 195, 597, 2019.
18. Akbari, H., Browne, M.C., Ortega, A., Huang, M.J., Hewitt, N.J., Norton, B., McCormack, S.J., Efficient energy storage technologies for photovoltaic systems. *Sol. Energy*, 192, 144, 2019.
19. Li, B., Roche, R., Paire, D., Miraoui, A., Sizing of a stand-alone microgrid considering electric power, cooling/heating, hydrogen loads and hydrogen storage degradation. *Appl. Energy*, 205, 1244, 2017.
20. Gu, W., Wu, Z., Bo, R., Liu, W., Zhou, G., Chen, W., Wu, Z., Modeling, planning and optimal energy management of combined cooling, heating and power microgrid: A review. *Int. J. Electr. Power Energy Syst.*, 54, 26, 2014.
21. Lehmann, C., Weeber, M., Böhner, J., Steinhilper, R., Techno-economical Analysis of Photovoltaic-battery Storage Systems for Peak-shaving Applications and Self-consumption Optimization in Existing Production Plants. *Procedia CIRP*, 48, 313, 2016.
22. Oudalov, A., Cherkaoui, R., Beguin, A., Sizing and optimal operation of battery energy storage system for peak shaving application. In: *2007 IEEE Lausanne Powertech, Proceedings*, pp. 621–625. IEEE, 2007.
23. Durbin, D.J., Malardier-Jugroot, C., Review of hydrogen storage techniques for on board vehicle applications. *Int. J. Hydrogen Energy*, 38, 14595, 2013.
24. Li, L., Xu, C., Chen, C., Wang, Y., Jiao, L., Yuan, H., Sodium alanate system for efficient hydrogen storage. *Int. J. Hydrogen Energy*, 38, 8798, 2013.
25. Muthukumar, P., Groll, M., Metal hydride based heating and cooling systems: A review. *Int. J. Hydrogen Energy*, 35, 3817, 2010.
26. Shiraz, H.G., Tavakoli, O., Investigation of graphene-based systems for hydrogen storage. *Renew. Sustain. Energy Rev.*, 74, 104, 2017.
27. Züttel, A., Materials for hydrogen storage. *Mater. Today*, 6, 24, 2003.

28. Yartys, V.A., Lototskyy, M. V., Akiba, E., Albert, R., Antonov, V.E., Ares, J.R., Baricco, M., Bourgeois, N., Buckley, C.E., Bellosta von Colbe, J.M., Crivello, J.C., Cuevas, F., Denys, R. V., Dornheim, M., Felderhoff, M., Grant, D.M., Hauback, B.C., Humphries, T.D., Jacob, I., Jensen, T.R., de Jongh, P.E., Joubert, J.M., Kuzovnikov, M.A., Latroche, M., Paskevicius, M., Pasquini, L., Popilevsky, L., Skripnyuk, V.M., Rabkin, E., Sofianos, M. V., Stuart, A., Walker, G., Wang, H., Webb, C.J., Zhu, M., Magnesium based materials for hydrogen based energy storage: Past, present and future. *Int. J. Hydrogen Energy*, 44, 7809, 2019.
29. Hardy, B., Tamburello, D., Corgnale, C., Hydrogen storage adsorbent systems acceptability envelope. *Int. J. Hydrogen Energy*, 43, 19528, 2018.
30. Muthukumar, P., Kumar, A., Raju, N.N., Malleswararao, K., Rahman, M.M., A critical review on design aspects and developmental status of metal hydride based thermal machines. *Int. J. Hydrogen Energy*, 43, 17753, 2018.
31. Pyle, D.S., Gray, E.M.A., Webb, C.J., Hydrogen storage in carbon nanostructures via spillover. *Int. J. Hydrogen Energy*, 41, 19098, 2016.
32. Callini, E., Aguey-Zinsou, K.F., Ahuja, R., Ares, J.R., Bals, S., Biliškov, N., Chakraborty, S., Charalambopoulou, G., Chaudhary, A.L., Cuevas, F., Dam, B., de Jongh, P., Dornheim, M., Filinchuk, Y., Grbović Novaković, J., Hirscher, M., Jensen, T.R., Jensen, P.B., Novaković, N., Lai, Q., Leardini, F., Gattia, D.M., Pasquini, L., Steriotis, T., Turner, S., Vegge, T., Züttel, A., Montone, A., Nanostructured materials for solid-state hydrogen storage: A review of the achievement of COST Action MP1103. *Int. J. Hydrogen Energy*, 41, 14404, 2016.
33. Liu, W., Webb, C.J., Gray, E.M.A., Review of hydrogen storage in AB3 alloys targeting stationary fuel cell applications. *Int. J. Hydrogen Energy*, 41, 3485, 2016.
34. Afzal, M., Mane, R., Sharma, P., Heat transfer techniques in metal hydride hydrogen storage: A review. *Int. J. Hydrogen Energy*, 42, 30661, 2017.
35. Lynch, F.E., Metal hydride practical applications. *J. Less-Common Met.*, 172–174, 943, 1991.
36. Muthukumar, P., Maiya, M.P., Murthy, S.S., Experiments on a metal hydride based hydrogen compressor. *Int. J. Hydrogen Energy*, 30, 879, 2005.
37. Muthukumar, P., Singh Patel, K., Sachan, P., Singhal, N., Computational study on metal hydride based three-stage hydrogen compressor. *Int. J. Hydrogen Energy*, 37, 3797, 2012.
38. Satya Sekhar, B., Muthukumar, P., Performance tests on a double-stage metal hydride based heat transformer. *Int. J. Hydrogen Energy*, 38, 15428, 2013.
39. Muthukumar, P., Satheesh, A., Analysis of crossed van't Hoff metal hydride based heat pump. *Int. J. Hydrogen Energy*, 38, 11415, 2013.
40. Satheesh, A., Muthukumar, P., Performance investigation of double-stage metal hydride based heat pump. *Appl. Therm. Eng.*, 30, 2698, 2010.
41. Satheesh, A., Muthukumar, P., Dewan, A., Computational study of metal hydride cooling system. *Int. J. Hydrogen Energy*, 34, 3164, 2009.

42. Muthukumar, P., Patil, M.S., Raju, N.N., Imran, M., Parametric investigations on compressor-driven metal hydride based cooling system. *Appl. Therm. Eng.*, 97, 87, 2016.
43. Balat, M., A review of modern wind turbine technology. *Energy Sources, Part A Recover. Util. Environ. Eff.*, 31, 1561, 2009.
44. Justice, S., The Genres of Piers Plowman. *Viator*, 19, 291, 1988.
45. Knight, J., Breezing into town. *Nature*, 430, 12, 2004.
46. Whitby, B., Ugalde-Loo, C.E., Performance of pitch and stall regulated tidal stream turbines. *IEEE Trans. Sustain. Energy*, 5, 64, 2014.
47. Roga, S., Bardhan, S., Kumar, Y., Dubey, S.K., Recent technology and challenges of wind energy generation: A review. *Sustain. Energy Technol. Assessments*, 52, 102239, 2022.
48. Chen, Z., Blaabjerg, F., Wind farm-A power source in future power systems. *Renew. Sustain. Energy Rev.*, 13, 1288, 2009.
49. Blanco, M.I., The economics of wind energy. *Renew. Sustain. Energy Rev.*, 13, 1372, 2009.
50. Korprasertsak, N., Leephakpreeda, T., Analysis and optimal design of wind boosters for Vertical Axis Wind Turbines at low wind speed. *J. Wind Eng. Ind. Aerodyn.*, 159, 9, 2016.
51. Njiri, J.G., Liu, Y., Söffker, D., Multivariable control of large variable-speed wind turbines for generator power regulation and load reduction. *IFAC-PapersOnLine*, 28, 544, 2015.
52. Onol, A.O., Yesilyurt, S., Effects of wind gusts on a vertical axis wind turbine with high solidity. *J. Wind Eng. Ind. Aerodyn.*, 162, 1, 2017.
53. Ackermann, T., Söder, L., An overview of wind energy-status 2002. *Renew. Sustain. Energy Rev.*, 6, 67, 2002.
54. Yang, H., Lu, L., Zhou, W., A novel optimization sizing model for hybrid solar-wind power generation system. *Sol. Energy*, 81, 76, 2007.
55. Yang, H.X., Lu, L., Burnett, J., Weather data and probability analysis of hybrid photovoltaic-wind power generation systems in Hong Kong. *Renew. Energy*, 28, 1813, 2003.
56. Liu, X., Lin, Z., Feng, Z., Short-term offshore wind speed forecast by seasonal ARIMA - A comparison against GRU and LSTM. *Energy*, 227, 120492, 2021.
57. Santhosh, M., Venkaiah, C., Vinod Kumar, D.M., Ensemble empirical mode decomposition based adaptive wavelet neural network method for wind speed prediction. *Energy Convers. Manag.*, 168, 482, 2018.
58. Ma, X., Jin, Y., Dong, Q., A generalized dynamic fuzzy neural network based on singular spectrum analysis optimized by brain storm optimization for short-term wind speed forecasting. *Appl. Soft Comput. J.*, 54, 296, 2017.
59. Zhou, W., Lou, C., Li, Z., Lu, L., Yang, H., Current status of research on optimum sizing of stand-alone hybrid solar-wind power generation systems. *Appl. Energy*, 87, 380, 2010.
60. Bajpai, P., Dash, V., Hybrid renewable energy systems for power generation in stand-alone applications: A review. *Renew. Sustain. Energy Rev.*, 16, 2926, 2012.

61. Craparo, E., Karatas, M., Singham, D.I., A robust optimization approach to hybrid microgrid operation using ensemble weather forecasts. *Appl. Energy*, 201, 135, 2017.
62. Abdin, Z., Mérida, W., Hybrid energy systems for off-grid power supply and hydrogen production based on renewable energy: A techno-economic analysis. *Energy Convers. Manag.*, 196, 1068, 2019.
63. Das, D.C., Roy, A.K., Sinha, N., GA based frequency controller for solar thermal-diesel-wind hybrid energy generation/energy storage system. *Int. J. Electr. Power Energy Syst.*, 43, 262, 2012.
64. Iverson, Z., Achuthan, A., Marzocca, P., Aidun, D., Optimal design of hybrid renewable energy systems (HRES) using hydrogen storage technology for data center applications. *Renew. Energy*, 52, 79, 2013.
65. Kelly, N.A., Hydrogen production by water electrolysis. In: *Advances in Hydrogen Production, Storage and Distribution*, pp. 159–185. Elsevier, 2014.
66. Ould Amrouche, S., Rekioua, D., Rekioua, T., Bacha, S., Overview of energy storage in renewable energy systems. *Int. J. Hydrogen Energy*, 41, 20914, 2016.
67. Ma, T., Yang, H., Lu, L., Peng, J., Technical feasibility study on a standalone hybrid solar-wind system with pumped hydro storage for a remote island in Hong Kong. *Renew. Energy*, 69, 7, 2014.
68. Giannakoudis, G., Papadopoulos, A.I., Seferlis, P., Voutetakis, S., Optimum design and operation under uncertainty of power systems using renewable energy sources and hydrogen storage. *Int. J. Hydrogen Energy*, 35, 872, 2010.
69. Yang, W.J., Aydin, O., Wind energy-hydrogen storage hybrid power generation. *Int. J. Energy Res.*, 25, 449, 2001.
70. Maleki, A., Design and optimization of autonomous solar-wind-reverse osmosis desalination systems coupling battery and hydrogen energy storage by an improved bee algorithm. *Desalination*, 435, 221, 2018.
71. Ishaq, H., Dincer, I., Crawford, C., A review on hydrogen production and utilization: Challenges and opportunities. *Int. J. Hydrogen Energy*, 47, 26238, 2022.
72. Wang, D., Muratori, M., Eichman, J., Wei, M., Saxena, S., Zhang, C., Quantifying the flexibility of hydrogen production systems to support large-scale renewable energy integration. *J. Power Sources*, 399, 383, 2018.
73. Sarker, A.K., Azad, A.K., Rasul, M.G., Doppalapudi, A.T., Prospect of Green Hydrogen Generation from Hybrid Renewable Energy Sources: A Review. *Energies*, 16, 1556, 2023.
74. Schrotenboer, A.H., Veenstra, A.A.T., uit het Broek, M.A.J., Ursavas, E., A Green Hydrogen Energy System: Optimal control strategies for integrated hydrogen storage and power generation with wind energy. *Renew. Sustain. Energy Rev.*, 168, 112744, 2022.
75. Fan, X. chao, Wang, W. qing, Shi, R. jing, Cheng, Z. jiang, Hybrid pluripotent coupling system with wind and photovoltaic-hydrogen energy storage and the coal chemical industry in Hami, Xinjiang. *Renew. Sustain. Energy Rev.*, 72, 950, 2017.

7

Technologies Based on Reusable Wind Turbine Blades

Abhinay Thakur[1] and Ashish Kumar[2*]

[1]Department of Chemistry, School of Chemical Engineering and Physical Sciences, Lovely Professional University, Phagwara, Punjab, India
[2]Nalanda College of Engineering, Bihar Engineering University, Science, Technology and Technical Education Department, Government of Bihar, Bihar, India

Abstract
The growing global demand for clean and sustainable energy sources has propelled the rapid development of wind power generation. Wind turbines, as the key components of this industry, require efficient and cost-effective solutions to maximize their potential. One critical aspect is the design and manufacturing of wind turbine blades, which significantly impact the overall performance and economic viability of wind farms. This chapter explores the emerging technologies centered around reusable wind turbine blades, highlighting their advancements, challenges, and implications. The chapter begins by providing an overview of conventional wind turbine blade materials and their limitations in terms of recyclability and environmental impact. It then delves into recent research and innovations in materials engineering, such as composite materials, bio-based resins, and additive manufacturing techniques. Furthermore, the chapter examines the challenges faced in implementing reusable blade technologies, including structural integrity, fatigue resistance, and manufacturing scalability.

Keywords: Wind power generation, reusable wind turbine blades, sustainable energy sources, materials engineering, recyclability, manufacturing scalability

**Corresponding author*: drashishchemlpu@gmail.com

Inamuddin, Tariq Altalhi and Mohammad Luqman (eds.) Wind Energy Storage and Conversion: From Basics to Utilities, (133–184) © 2024 Scrivener Publishing LLC

7.1 Introduction

The increasing global demand for clean and sustainable energy sources is driven by the urgent need to address climate change and reduce our reliance on fossil fuels, which are major contributors to greenhouse gas emissions [1, 2]. As a result, renewable energy technologies, including wind power generation, have gained significant attention and investment worldwide. Wind turbines are at the forefront of the wind power industry, responsible for converting the kinetic energy of wind into electricity. They consist of several crucial components, but one of the most vital is the design and manufacturing of wind turbine blades [3, 4]. These blades are essential for capturing the energy from the wind and transferring it to the turbine's rotor, where it is converted into electrical energy. The efficiency and performance of wind turbines greatly depend on the design and quality of their blades. The design of wind turbine blades must take into account various factors, including aerodynamics, structural integrity, and material selection. Efficient blade design ensures optimal energy capture, enabling wind turbines to generate electricity at maximum capacity [5–7]. Moreover, the manufacturing process plays a crucial role in ensuring the reliability, durability, and cost-effectiveness of the blades. Figure 7.1 illustrates the installation process of a blade at Trianel Windpark Borkum II.

Figure 7.1 Blade installation taking place at Trianel Windpark Borkum II. Adapted with permission from [91] MDPI. Distributed under Creative Commons attribution-based license (CCBY 4.0).

With the growing demand for wind energy, there has been a continuous trend toward larger and more powerful wind turbines. Longer blades are necessary to capture more wind energy and increase the electricity generation capacity of wind farms. However, this trend presents unique challenges in blade design and manufacturing. One of the key challenges is balancing the structural integrity and weight of the blades. Longer blades experience greater stresses and loads, requiring careful design considerations to ensure that they can withstand harsh wind conditions and operate reliably over their lifespan [6, 8, 9]. Additionally, the choice of materials for blade construction is crucial. Traditional wind turbine blades have been predominantly made of fiberglass-reinforced composites, which offer desirable properties such as strength and flexibility. However, these materials have limitations in terms of recyclability and sustainability, leading to environmental concerns. Following are several global examples highlighting the challenges and initiatives associated with wind turbine blades and their environmental impact:

United States: The United States has been a leader in wind power generation, with a significant number of wind farms across the country. However, the disposal of decommissioned wind turbine blades has become a concern. According to the National Renewable Energy Laboratory (NREL), around 8,000 blades are expected to be retired annually by 2024, leading to approximately 1.5 million tons of blade waste [10]. To address this issue, researchers and industry stakeholders are actively exploring recycling technologies and alternative blade materials to minimize waste and improve sustainability.

Europe: Europe has been at the forefront of renewable energy adoption, with countries like Germany, Denmark, and the Netherlands investing heavily in wind power. As wind turbines reach the end of their operational life, European countries are facing challenges in managing the growing volume of decommissioned blades. Several initiatives have been launched to tackle this issue, such as the Re-Wind project in Denmark, which aims to develop sustainable recycling methods for wind turbine blades, and the Horizon 2020-funded project "BladeTrack," which focuses on the tracking and recycling of wind turbine blades [11–13].

China: China has experienced a significant growth in wind power capacity, making it the world's largest market for wind energy. However, the disposal of wind turbine blades has become a concern due to their non-biodegradable nature. To address this issue, Chinese researchers and industry players have been exploring various solutions—for example, a research team at the Beijing University of Technology has developed a method for converting decommissioned wind turbine blades into construction materials like cement and concrete, reducing waste and environmental impact.

Australia: Australia has vast wind resources and has been increasing its wind power capacity in recent years. With the expansion of wind farms, there is a growing focus on sustainable practices and minimizing the environmental impact of wind turbine blades. The Australian Renewable Energy Agency (ARENA) has supported research projects to investigate innovative blade materials and recycling techniques [14–16]. Additionally, partnerships between industry and research institutions aim to develop sustainable blade designs and explore options for repurposing retired blades.

Brazil: Brazil has made significant strides in wind power development, particularly in its northeastern region. As wind farms mature and older turbines are replaced, the disposal and recycling of wind turbine blades have become important considerations. The Brazilian Wind Energy Association (ABEEólica) is actively promoting research and collaboration to address the environmental challenges associated with blade waste and foster sustainable practices in the wind energy sector.

These global examples highlight the urgency and ongoing efforts to find effective solutions for the design, manufacturing, and end-of-life management of wind turbine blades. Governments, research institutions, and industry stakeholders are collaborating to develop technologies and strategies that promote the reuse, recycling, and sustainable disposal of wind turbine blades, aiming to maximize the benefits of wind power generation while minimizing its environmental impact. The continuous growth of wind energy has led to the development of larger and more powerful wind turbines, with longer blades that can capture greater amounts of wind energy. However, the design and manufacturing of these blades present unique challenges, particularly in terms of their environmental impact and end-of-life management [17, 18]. Traditional wind turbine blades are predominantly made from fiberglass-reinforced composites, which offer favorable mechanical properties but pose limitations in terms of recyclability and sustainability. As the number of decommissioned wind turbine blades increases, finding effective solutions for their disposal or recycling becomes a pressing concern.

In response to these challenges, there is a growing interest in technologies centered around reusable wind turbine blades. Reusable blades aim to address the environmental and economic implications of conventional blade materials by incorporating innovative designs, materials, and manufacturing techniques that enhance recyclability, reusability, and overall sustainability. These technologies have the potential to revolutionize the wind energy industry, offering efficient and cost-effective solutions that maximize energy production and minimize the environmental footprint of wind farms. The objectives of this chapter are to explore the advancements, challenges, and implications of technologies based on reusable wind turbine blades. By examining the latest research and innovations in materials engineering, manufacturing processes, and design optimization, this chapter aims to provide a comprehensive overview of the state-of-the-art approaches for developing reusable wind turbine blades. Furthermore, the chapter will discuss the challenges faced in implementing these technologies, such as ensuring structural integrity, fatigue resistance, and manufacturing scalability.

One key aspect to be addressed in this chapter is the limitations of conventional wind turbine blade materials in terms of recyclability and environmental impact. The use of fiberglass-reinforced composites, while providing desirable mechanical properties, poses challenges in terms of end-of-life disposal and sustainability. Alternative materials and composites, including bio-based resins and recycled materials, will be explored as potential solutions to enhance recyclability and reduce the environmental impact of wind turbine blades. Additionally, the chapter will delve into the emerging advancements in materials engineering for reusable blades. The utilization of composite materials, such as carbon fiber-reinforced polymers, as well as the integration of bio-based resins derived from renewable sources, has shown promise in enhancing the sustainability and reusability of wind turbine blades. Furthermore, additive manufacturing techniques, including 3D printing, offer new possibilities for the production of complex blade designs and customization, contributing to increased efficiency and reduced waste.

7.2 Wind Power Generation and the Importance of Wind Turbine Blades

7.2.1 Global Demand for Clean and Sustainable Energy

The increasing global demand for clean and sustainable energy sources stems from the urgent need to address climate change and reduce our reliance on fossil fuels. The burning of fossil fuels for energy production releases greenhouse gases into the atmosphere, contributing to the greenhouse effect and global warming. This has led to adverse effects such as rising sea levels, extreme weather events, and the loss of biodiversity [19]. To mitigate these impacts and transition to a more sustainable future, the global community has recognized the importance of adopting renewable energy sources. Wind power generation has become a central focus in the quest for sustainable energy solutions. Wind turbines harness the kinetic energy of the wind and convert it into electricity through the rotation of a rotor. They offer numerous advantages over conventional energy sources. Firstly, wind power is a clean source of energy, as it does not produce harmful emissions or pollutants during operation. This helps to improve air quality and reduce the environmental impact associated with traditional energy generation. Secondly, wind is a renewable resource that is abundant and widely available. Unlike finite fossil fuel resources, wind energy can be harnessed indefinitely, making it a sustainable and long-term solution for meeting our energy needs.

The global demand for wind power has seen significant growth in recent years. According to the Global Wind Energy Council (GWEC), the cumulative installed capacity of wind power reached 743 gigawatts (GW) by the end of 2020, representing a substantial increase from previous years [16, 17, 20, 21]. Countries around the world, including China, the United States, Germany, India, and Spain, have been leading the way in wind power installations. This trend is expected to continue as more countries prioritize renewable energy and set ambitious targets for increasing their share of wind power in the energy mix. One of the key components of a wind turbine that plays a critical role in wind power generation is the wind turbine blade. The design and manufacturing of wind turbine blades have a significant impact on the overall performance and economic viability of wind farms. The blades are responsible for capturing the energy from the wind and converting it into rotational motion. The efficiency and effectiveness of this energy conversion process depend on the design, materials, and construction of the blades. As the wind energy industry continues to expand, there has been a trend towards larger and more powerful wind turbines, necessitating longer blades capable of capturing greater amounts of wind energy. Longer blades have a larger swept area, enabling them to intercept more wind and generate higher electricity output [22, 23]. Moreover, the aerodynamic design of the blades plays a crucial role in optimizing energy conversion and minimizing energy losses due to turbulence or drag. A well-designed blade shape can enhance the efficiency and performance of the turbine, maximizing its power output.

The structural integrity of wind turbine blades is another crucial aspect to consider. The blades are subjected to various dynamic forces, including wind gusts, turbulence, and vibrations. They must be able to withstand these loads and maintain their integrity over the lifespan of the turbine. Failure or damage to the blades can lead to reduced energy output, increased maintenance costs, and potential safety hazards. Therefore, the materials and construction techniques used for blade manufacturing must provide sufficient strength, stiffness, and fatigue resistance. Traditionally, wind turbine blades have been constructed using fiberglass-reinforced composites. These materials offer favorable mechanical properties, such as high strength-to-weight ratio and corrosion resistance. However, they present limitations in terms of recyclability and environmental sustainability. As the number of decommissioned wind turbine blades increases, finding effective solutions for their disposal or recycling becomes a pressing concern. Improper disposal can lead to environmental pollution and waste management challenges. To address these challenges, there is ongoing research and innovation focused on developing technologies and materials for reusable wind turbine blades [24, 25]. Advanced composite materials, such as carbon fiber-reinforced polymers, are being explored for their superior mechanical properties, including higher strength and stiffness. These materials can enable the production of longer and lighter blades, improving energy capture and reducing material usage. Additionally, bio-based resins derived from renewable sources are being investigated as greener alternatives to conventional blade materials, offering improved recyclability and reduced environmental impact.

7.2.2 Role of Wind Turbines in Wind Power Generation

Wind turbines play a pivotal role in wind power generation by harnessing the renewable energy of the wind and converting it into electricity [26, 27]. They are the primary means by which wind energy is captured and transformed into a usable form for powering homes, businesses, and industries. The basic structure of a wind turbine consists of several essential components. At the base of the turbine is the tower, which provides support and elevation for the other components. The tower is designed to withstand the various forces exerted by the wind and to ensure the stability of the entire structure. Towers can vary in height depending on the size and type of wind turbine, with taller towers enabling access to stronger and more consistent wind resources at higher altitudes. Above the tower, the nacelle is positioned. The nacelle houses the crucial machinery and equipment required for the generation of electricity. This includes the generator, which converts the mechanical energy from the rotor into electrical energy. Additionally, the nacelle contains other components such as the gearbox, brake system, control systems, and sensors, all of which are necessary for the efficient operation and control of the turbine [28, 29].

The rotor, located at the front of the turbine, consists of the hub and the blades. The hub is connected to the main shaft within the nacelle, and it serves as the central connection point for the blades. The blades, typically three in number, are designed to capture the kinetic energy of the wind as it flows past them. They are aerodynamically shaped to maximize the amount of energy extracted from the wind, allowing for the most efficient conversion into rotational motion. The length and design of the wind turbine blades play a crucial role in determining the turbine's performance. Longer blades have a larger swept area, which means they can capture more wind and generate greater rotational energy. This translates into higher power output and increased electricity generation. Therefore, advancements in blade design, including increased length and improved aerodynamics, have significantly contributed to the overall efficiency and viability of wind turbines. The captured rotational energy from the blades is transferred to the generator within the nacelle through the main shaft and gearbox. The generator converts the mechanical energy into electrical energy by employing the principle of electromagnetic induction. As the rotor spins, the magnetic field within the generator is manipulated, inducing a flow of electric current in the windings. This current is then collected and transmitted through electrical systems to be utilized or fed into the power grid for distribution.

The electricity generated by wind turbines is typically in the form of alternating current (AC). To make it compatible with the electrical grid, it is often converted to a higher voltage and transmitted over long distances through power lines [30, 31]. Once it reaches its destination, the electricity can be used directly by consumers or businesses, or it can be stored for later use in batteries or other energy storage systems. Wind turbines have a significant advantage in that wind is a renewable and widely available resource. Unlike fossil fuels, wind energy does not deplete or contribute to harmful greenhouse gas emissions when used for electricity generation. Wind power is a clean and sustainable alternative, reducing reliance on fossil fuels and helping to combat climate change. The role of wind turbines in wind power generation extends beyond their immediate function of converting wind energy into electricity. They also contribute to the diversification of energy sources, reduce greenhouse gas emissions, create jobs in the renewable energy sector, and promote energy independence. As countries around the world strive to increase their share of renewable energy, wind turbines and the continued advancement of their technology play a crucial role in achieving a sustainable and low-carbon future.

7.2.3 Impact of Wind Turbine Blades on Performance and Viability

The design and manufacturing of wind turbine blades have a significant impact on the performance and viability of wind power generation. As the primary components responsible for capturing the kinetic energy of the wind, the efficiency, structural integrity, and material properties of the blades directly influence the power output, reliability, and cost-effectiveness of wind turbines. The efficient blade design is crucial for maximizing the energy conversion from the wind [30–32]. The length, shape, and curvature of the blades are carefully optimized to achieve the highest possible energy capture. Longer blades have a larger swept area, allowing them to intercept a greater volume of wind and extract more energy. Additionally, the aerodynamic profile of the blades is designed to minimize drag and turbulence, enabling smoother airflow and improved efficiency. By enhancing the aerodynamic performance of the blades, the overall efficiency of wind turbines can be significantly increased, leading to higher electricity generation. The structural integrity of wind turbine blades is essential to ensure their performance and longevity. The blades must withstand the dynamic forces exerted by the wind, including high wind speeds, gusts, and turbulent conditions. Failure or damage to the blades can result in reduced energy output, increased maintenance costs, and potentially catastrophic failures. Therefore, the materials and construction techniques used for blade manufacturing must provide sufficient strength, stiffness, and fatigue resistance.

Traditionally, wind turbine blades have been predominantly made of fiberglass-reinforced composites. These materials offer a favorable combination of strength, weight, and cost-effectiveness. However, they pose challenges in terms of recyclability and sustainability. With the growing number of decommissioned wind turbine blades, finding effective solutions for their disposal or recycling has become a pressing concern. The development of advanced composite materials, such as carbon fiber-reinforced polymers, is being explored to improve the mechanical properties of the blades while also enhancing their recyclability. Moreover, the choice of materials for wind turbine blades has implications for their environmental impact throughout their life cycle. From raw material extraction to manufacturing, transportation, operation, and eventual decommissioning, the environmental footprint of the blades needs to be considered. Efforts are being made to develop more sustainable materials, such as bio-based resins derived from renewable sources. These materials offer the potential to reduce the environmental impact of wind turbine blades by utilizing renewable resources and reducing reliance on fossil fuel-derived materials.

The manufacturing processes for wind turbine blades are also evolving to improve performance and reduce costs. Advanced manufacturing techniques, such as automated manufacturing and robotic systems, are being employed to enhance precision and efficiency in blade production. Furthermore, additive manufacturing, commonly known as 3D printing, is being explored as a means to optimize blade designs and reduce material waste [33–35]. This technology allows for complex geometries and customization, enabling the production of blades that are specifically tailored to different wind conditions and turbine specifications. The impact of wind turbine blades extends beyond their immediate function of converting wind energy into electricity. The reliability and maintenance requirements of the blades significantly affect the economic viability of wind projects. Blades that are prone to damage or require frequent maintenance can lead to increased downtime and higher operational costs. Therefore, advancements in blade design and materials that enhance durability and reduce maintenance needs are crucial for maximizing the economic benefits of wind power generation.

Furthermore, the deployment of wind farms is often subject to regulatory and environmental considerations. The visual impact of wind turbines, including their blades, on the landscape and local communities is an important aspect to address. Efforts are being made to design blades that are not only efficient but also visually appealing, minimizing visual footprint and enhancing the public acceptance of wind energy projects.

7.3 Conventional Wind Turbine Blade Materials and Limitations

7.3.1 Overview of Conventional Blade Materials

Conventional wind turbine blades rely on the use of fiberglass-reinforced composites due to their excellent combination of mechanical properties [36–38]. Fiberglass, also known as glass fiber, is a lightweight and high-strength material that is commonly used in various industries, including aerospace, automotive, and construction. In wind turbine blade applications, fiberglass provides the necessary strength and stiffness to withstand the dynamic forces exerted by the wind. The manufacturing process for conventional wind turbine blades begins with the creation of a mold. The mold is typically made of two halves, representing the positive and negative shapes of the blade. These molds are usually constructed using materials such as fiberglass or steel. Layers of fiberglass fabric are then laid inside the mold. The fiberglass fabric consists of continuous strands of glass fibers, which are arranged in a specific orientation to optimize the strength and performance of the blade. The fabric is cut and placed in a way that follows the desired blade curvature and shape [39].

Next, the resin matrix is applied to impregnate the fiberglass fabric. The resin matrix serves as the binding material that holds the glass fibers together, providing structural integrity to the composite. Epoxy and polyester resins are commonly used in wind turbine blade manufacturing due to their excellent adhesion to the glass fibers and their ability to cure into a rigid solid. The impregnated fiberglass fabric is carefully layered inside the mold, ensuring that the fibers are evenly distributed and there are no voids or air bubbles [40, 41]. This process is often done manually or using automated equipment to ensure precision and consistency. Once the fiberglass fabric is in place, the mold is closed, and the blade is subjected to a curing process. Curing involves applying heat and pressure to the blade, allowing the resin to chemically react and harden. The curing process typically takes several hours or even days, depending on the specific resin system and curing conditions. After curing, the mold is opened, and the blade is carefully removed. Any excess material or imperfections are trimmed or sanded to achieve the final desired shape and finish. The blade is then inspected for quality assurance, ensuring that it meets the required specifications and standards.

The use of fiberglass-reinforced composites in wind turbine blades offers several advantages. Firstly, fiberglass has a high strength-to-weight ratio, making the blades lightweight yet capable of withstanding the mechanical loads imposed by the wind. This lightweight characteristic allows for easier installation and reduces the structural demands on the entire turbine system. Additionally, fiberglass composites offer excellent fatigue resistance, which is essential for wind turbine blades that continuously experience cyclic loading due to wind gusts and turbulence. The ability of the material to withstand repeated stress cycles without significant degradation is crucial for the long-term performance and reliability of the blades. Moreover, fiberglass-reinforced composites provide good corrosion resistance, making the blades suitable for use in harsh environments, including offshore wind farms. The non-metallic nature of fiberglass eliminates the risk of corrosion, reducing maintenance requirements and prolonging the operational life of the blades. Pathak *et al.* [42] presented a novel approach using cellulose microparticles (CMP) and cellulose microfibers (CMF) as fillers to reinforce water-based polyurethane (PU) coatings on glass fiber-reinforced polymer (GFRP) substrates. These coatings were developed using a simple spray method. Field emission scanning electron microscopy images reveal the agglomerated nature of CMP particles and the fibrous morphology of CMF. Fourier transform infrared spectra demonstrate the presence of cellulose and polyurethane through the characteristic absorption bands of C–H, C=O, and N–H. Thermal gravimetric analysis results reveal the stability of CMP up to 285°C, while CMF degradation occurs at 243°C. X-ray photoelectron spectroscopy illustrates the presence of C–C/C–H, C–O, C–OH, and O–C=O bonds associated with cellulose structure in CMP, CMF, and the corresponding coatings. The coatings' resistance to solid particle erosion was evaluated at different concentrations of CMP and CMF, using impact angles of 30° and 90°. The erosion scars of PUCMF 5 wt.% coatings at 30° and 90° impact angles are depicted in Figure 7.2a and Figure 7.2b, respectively, displaying their characteristic 3D profiles. The profiles reveal that the erosion scar shape for a 30° impact angle resembles an elliptical cone, whereas for a 90° impact angle, it resembles a circular cone. It is worth noting that in soft polymeric coatings, erodent particles may become embedded within the coating during the solid particle erosion (SPE) test, leading to weight gain after testing. Therefore, in this study,

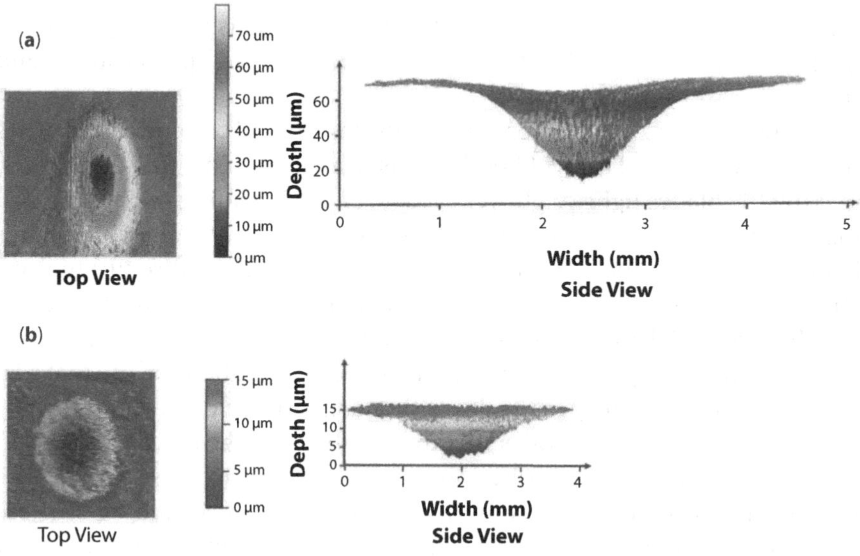

Figure 7.2 3D top views and side views of the erosion scars of the PUCMF 5 wt.% coating. (a) Erosion scar at 30° impact angle. (b) Erosion scar at 90° impact angle. Adapted with permission from [42] MDPI. Distributed under Creative Commons attribution-based license (CCBY 4.0).

material loss was evaluated based on volume loss rather than weight loss. The volume losses were determined by analyzing the 2D profiles of the erosion scars. Among the coatings, the CMP-reinforced PU coating with a 5 wt.% filler concentration and the CMF-reinforced PU coating with a 2 wt.% concentration exhibited the best performance against solid particle erosion. Additionally, nanoindentation studies were conducted to establish a relationship between H3/E2 and the average erosion rate of the coatings.

However, there are also limitations associated with the use of conventional blade materials. One significant limitation is the recyclability of fiberglass-reinforced composites. Due to the complex nature of the material, separating and recovering the different components during the recycling process can be challenging. As a result, most decommissioned wind turbine blades end up in landfills, contributing to waste accumulation and environmental concerns. Another limitation is the environmental impact associated with the manufacturing process of fiberglass composites. The production of fiberglass and the resin matrix requires energy-intensive processes and the use of chemicals, which can contribute to carbon emissions and environmental pollution if not properly managed. Additionally, the disposal of wind turbine blades at the end of their life cycle raises concerns about potential environmental contamination due to the presence of resins and other components. To address these limitations, ongoing research and development efforts are focused on the exploration of alternative blade materials and manufacturing techniques. Advanced composite materials, such as carbon fiber-reinforced polymers (CFRP), are being investigated for their superior mechanical properties and recyclability potential. Bio-based resins derived from renewable resources are also being explored as greener alternatives to traditional petroleum-based resins.

7.3.2 Limitations in Terms of Recyclability and Environmental Impact

While conventional wind turbine blades made from fiberglass-reinforced composites offer numerous advantages in terms of mechanical properties and performance, they also pose significant limitations in terms of recyclability and environmental impact. These limitations have led to increasing concerns regarding the end-of-life management of wind turbine blades and the sustainability of the wind energy industry as a whole [43–45]. One of the major challenges associated with fiberglass-reinforced composites is their limited recyclability. The complex composition of these materials, consisting of a combination of glass fibers and resin matrix, makes it difficult to separate and recover the different components during the recycling process. The fiberglass fibers are durable and have a long lifespan, while the resin matrix is thermosetting and cannot be easily re-melted and reshaped. As a result, the recycling of wind turbine blades becomes a complicated and costly process.

Currently, the most common approach for recycling wind turbine blades is through mechanical grinding or shredding, which reduces the blades to small fragments. However, this process only allows for the recovery of the fiberglass fibers, while the resin matrix is typically discarded as waste. The recovered fibers can be used in low-value applications, such as reinforcing concrete or producing insulation materials. However, the recycling rate is still relatively low, and the overall environmental impact of the recycling process remains a concern. In addition to the limited recyclability, the manufacturing process of fiberglass composites also has environmental implications [46–48]. The production of fiberglass requires energy-intensive processes, such as the melting and spinning of glass fibers, which contribute to carbon emissions and energy consumption. Similarly, the production of the resin matrix involves the use of chemicals and petroleum-based raw materials, further contributing to environmental pollution and resource depletion. Furthermore, the disposal of wind turbine blades at the end of their operational life presents a significant environmental challenge. As the number of decommissioned wind turbine blades continues to increase, finding sustainable and responsible solutions for their disposal becomes imperative. Currently, a significant portion of decommissioned blades ends up in landfills, which not only consumes valuable landfill space but also poses a risk of potential environmental contamination due to the presence of resins and other components.

To address these limitations and enhance the sustainability of wind turbine blades, ongoing research and innovation are focused on developing alternative materials and manufacturing processes. One promising avenue is the exploration of advanced composite materials, such as carbon fiber-reinforced polymers (CFRP), which offer superior mechanical properties and improved recyclability compared to fiberglass composites [19, 49, 50]. CFRP blades have the potential for higher recycling rates, as carbon fibers can be recovered and reused in various applications. Another approach is the investigation of bio-based resins derived from renewable resources as a greener alternative to petroleum-based resins. Bio-based resins have the advantage of being more environmentally friendly in terms of their production and disposal. They can reduce the carbon footprint associated with the manufacturing process and offer improved end-of-life options, such as biodegradation or composting. Furthermore, research efforts are also directed toward the development of innovative recycling technologies specifically tailored for wind turbine blades. These technologies aim to overcome the challenges associated with the separation and recovery of different materials, enabling a higher recycling rate and reducing the environmental impact of blade disposal.

7.4 Advancements in Materials Engineering for Reusable Wind Turbine Blades

7.4.1 Composite Materials in Blade Design

Composite materials have revolutionized the design and manufacturing of wind turbine blades, with carbon fiber-reinforced polymers (CFRP) emerging as a promising alternative to conventional fiberglass-reinforced composites. CFRP blades utilize carbon fibers as the reinforcing element, which are embedded in a polymer matrix [14, 51]. The exceptional properties of carbon fibers make them highly suitable for wind turbine blade applications. Carbon fibers have an incredibly high strength-to-weight ratio, providing excellent structural integrity while minimizing weight. This enables the production of longer blades that can capture more wind energy and generate higher power output. The stiffness of carbon fibers also contributes to the overall stability and performance of the blades.

By utilizing CFRP materials, wind turbine manufacturers can achieve significant improvements in blade efficiency and energy capture. The lightweight nature of carbon fiber allows for reduced loads on the turbine, enabling it to operate more effectively even in low wind conditions. This results in higher annual energy production and improved overall performance of wind farms. Furthermore, the recyclability of CFRP materials presents a significant advantage over traditional fiberglass composites. As wind turbine blades reach the end of their operational life, the recycling of composite materials becomes crucial to minimize waste and environmental impact. Carbon fibers can be recovered from decommissioned blades through various recycling processes, such as mechanical or chemical methods. These fibers can then be reused in other applications, reducing the demand for virgin carbon fibers and conserving valuable resources. Li *et al.* [8] synthesized a biochar/polypyrrole coating on the surface of biochar. Characterization of the coating revealed its porous structure, irregular dents, and bumps, which contributed to the formation of a nanoscale roughness structure with excellent super-hydrophobic properties. Figure 7.3a illustrates a low-speed, return-flow wind tunnel equipped with

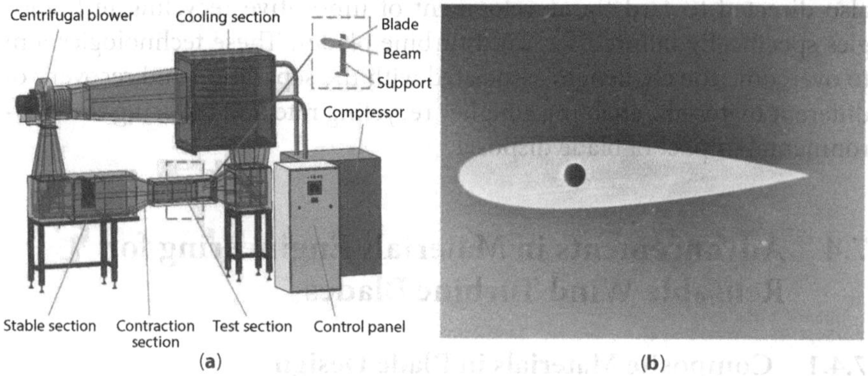

Figure 7.3 Schematic diagram depicting (a) the icing wind tunnel test system and (b) the test blade model. Adapted with permission from [8] MDPI. Distributed under Creative Common attribution-based license (CCBY 4.0).

refrigeration and spray devices. These devices were incorporated into conventional wind tunnels to generate a low-temperature icing environment.

The coating also exhibited a sufficient surface area. The wetting characteristics of the coating were evaluated using contact angle measurements, which showed a contact angle of 151°, indicating the coating's exceptional hydrophobicity. Icing wind tunnel tests were conducted to assess the anti-icing performance of the biochar coating and the biochar/polypyrrole coating at different ambient temperatures and wind speeds. To analyze the morphology and structure of the coatings prepared in this study, scanning electron microscopy (SEM) was employed. The experimental voltage was set at 5 kV. Figures 7.4a and 7.4b display the microstructure of biochar and biochar/polypyrrole, respectively. The image in Figure 7.4a reveals that the biochar material exhibits a highly porous structure. On the other hand, the scan of the biochar/polypyrrole in Figure 7.4b shows a honeycomb-like structure. Upon closer examination, it is evident that the biochar surface is adorned with polypyrrole nanowires, which display a uniform morphology and have micrometer-level lengths. These porous structures, as depicted in Figures 7.4a and 7.4b, contribute to a large specific surface area and rough surface characteristics, thereby enhancing the hydrophobicity of the coatings. The elemental composition of the biochar, as indicated by the EDS analysis diagram in Figure 7.4c, is primarily composed of carbon (C) and oxygen (O) elements. Similarly, the EDS analysis diagram of biochar/polypyrrole in Figure 7.4d reveals that the elemental composition consists of carbon (C), oxygen (O), and nitrogen (N). Furthermore, the biochar/polypyrrole coating exhibited the most significant anti-icing effect. This study

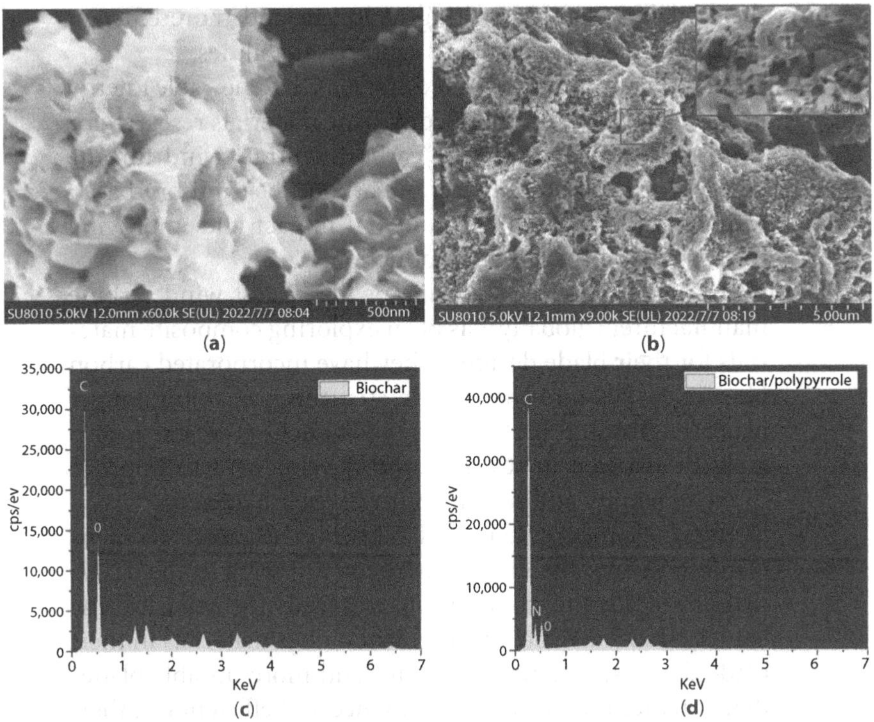

Figure 7.4 SEM images of (a) biochar, (b) biochar/polypyrrole, EDS analysis, (c) biochar, and (d) biochar/polypyrrole. Adapted with permission from [8] MDPI. Distributed under Creative Commons attribution-based license (CCBY 4.0).

provides practical insights for the development of anti-icing coatings for wind turbine blades.

Advancements in recycling technologies for CFRP are continuously being explored to enhance the efficiency and viability of the recycling process. Methods such as pyrolysis, solvent-based extraction, and mechanical grinding have shown promise in separating and recovering carbon fibers from the polymer matrix. These recycled fibers can be incorporated into new composite materials or utilized in other industries, such as automotive, aerospace, or construction. Some global examples of advancements in composite materials for wind turbine blades are as follows:

- LM Wind Power (Denmark): LM Wind Power, a leading manufacturer of wind turbine blades, has made significant progress in the development of composite materials for blades. They have successfully integrated carbon fiber into their blade designs, resulting in longer and more efficient blades with increased energy capture capabilities.

- GE Renewable Energy (United States): GE Renewable Energy has been at the forefront of composite materials research for wind turbine blades. Their "carbon-neutral" wind turbine blades utilize a combination of carbon fiber and bio-based resins, reducing the environmental impact and improving the sustainability of blade production.
- Vestas (Denmark): Vestas, one of the largest wind turbine manufacturers globally, has been exploring composite materials for their blade designs. They have incorporated carbon fiber materials into their blade structures, enabling longer blade lengths and higher energy production. Vestas is also actively involved in research and development to improve the recyclability and sustainability of their blades.
- Siemens Gamesa Renewable Energy (Spain): Siemens Gamesa Renewable Energy has been investing in the development of advanced composite materials for wind turbine blades. They have introduced carbon fiber materials in their blade designs, resulting in lighter and more durable blades that enhance the overall performance and efficiency of their turbines.
- Goldwind (China): Goldwind, a major wind turbine manufacturer in China, has been focusing on the use of advanced composite materials for their blades. They have integrated carbon fiber into their blade designs, leading to improved strength, stability, and energy capture capabilities.

However, there are challenges associated with the widespread adoption of CFRP in wind turbine blades. The cost of carbon fibers is higher compared to fiberglass, making CFRP blades more expensive to manufacture. This cost factor poses a barrier to the widespread implementation of CFRP blades, particularly in large-scale wind farms. Additionally, the production process for CFRP blades requires specialized manufacturing techniques and expertise, adding to the complexity and cost of blade production.

7.4.2 Bio-Based Resins for Sustainable Blades

In the quest for more sustainable wind turbine blades, there has been a growing interest in the development and utilization of bio-based resins. Bio-based resins are derived from renewable sources such as plant-based materials, offering a greener alternative to conventional petroleum-based resins. These resins have the potential to significantly reduce the environmental impact of blade manufacturing and improve the overall sustainability of wind power generation. One of the primary advantages of bio-based resins is their lower carbon footprint. Petroleum-based resins are derived from non-renewable fossil fuels and contribute to greenhouse gas emissions throughout their production process. In contrast, bio-based resins are derived from renewable resources such as vegetable oils, starches, or lignocellulosic biomass. The production of bio-based resins generally requires less energy and generates fewer greenhouse gas emissions, making them more environmentally friendly. Furthermore, bio-based resins offer the potential for improved end-of-life management of wind turbine blades. Traditional fiberglass blades pose challenges in terms of their recyclability and disposal [15, 17, 52]. Since fiberglass composites are not easily recyclable, decommissioned blades often end up in landfills, contributing to waste accumulation. In contrast, bio-based resins can be more easily decomposed or recycled, offering a viable solution for reducing waste and promoting a circular economy.

The use of bio-based resins in blade manufacturing also presents opportunities for local sourcing and regional economic development. Many bio-based resins can be derived from locally available agricultural or forestry by-products, creating opportunities for rural communities to participate in the renewable energy supply chain. This localized sourcing of materials can enhance regional economic growth and job creation and can reduce dependence on imported resources. Some global examples of the use of bio-based resins in wind turbine blade manufacturing are as follows:

- Vestas: Vestas, a leading wind turbine manufacturer, has been actively researching and developing bio-based resins for blade production. In collaboration with partners, they have successfully incorporated bio-based resins derived from vegetable oils into their blade manufacturing process. These bio-based resins offer improved recyclability and reduced environmental impact compared to traditional resin systems.
- LM Wind Power: LM Wind Power, a subsidiary of GE Renewable Energy, has been working on sustainable blade solutions, including the use of bio-based resins. They have developed a prototype blade using a bio-based epoxy resin derived from vegetable oils. The prototype blade demonstrated comparable performance to conventional blades

while reducing the carbon footprint associated with resin production.

- Siemens Gamesa Renewable Energy: Siemens Gamesa Renewable Energy has also explored the use of bio-based resins in wind turbine blades. They have successfully produced blades using bio-based epoxy resins derived from vegetable oils. These bio-based resins offer advantages in terms of recyclability and reduced energy consumption during the manufacturing process [53, 54].
- Enercon: Enercon, a German wind turbine manufacturer, has been actively researching sustainable materials for wind turbine blade production. They have explored the use of bio-based resins derived from renewable sources such as vegetable oils and plant-based polymers. These bio-based resins offer improved environmental performance and contribute to the overall sustainability of wind energy.
- Research institutions: Several research institutions and universities worldwide are also investigating the potential of bio-based resins for wind turbine blades—for example, the Technical University of Denmark (DTU) has conducted research on bio-based resins derived from plant-based materials for blade manufacturing. These studies aim to develop sustainable and recyclable solutions for the wind energy industry [16, 21, 23, 55].

However, the widespread adoption of bio-based resins for wind turbine blades still faces certain challenges. One key challenge is the need for cost-effective and scalable production processes. Bio-based resins are often more expensive to produce than petroleum-based resins, mainly due to the higher cost of feedstock and processing technologies. Scaling up production and optimizing manufacturing processes are essential to reduce costs and make bio-based resins economically competitive. Another challenge is ensuring the performance and durability of wind turbine blades made with bio-based resins. The mechanical properties of bio-based resins may differ from those of petroleum-based resins, requiring careful formulation and optimization to meet the stringent performance requirements of wind turbine blades [56–65]. Research and development efforts are focused on enhancing mechanical properties, such as strength, stiffness, and fatigue resistance, to ensure the long-term reliability and performance of blades made with bio-based resins. Several companies and research institutions are actively exploring the use of bio-based resins in wind turbine blades—for example, GE Renewable Energy has developed a prototype blade using bio-based resins derived from vegetable oils. The prototype demonstrated comparable performance to traditional blades while reducing the environmental impact.

7.4.3 Additive Manufacturing Techniques for Blade Production

Additive manufacturing, also known as 3D printing, has emerged as a disruptive technology in various industries, including the manufacturing of wind turbine blades. This innovative approach offers new possibilities for designing and producing blades with increased efficiency, reduced costs, and improved sustainability. Additive manufacturing techniques have the potential to revolutionize the wind energy industry by enabling the production of complex blade geometries, customization, and reduced material waste. One of the key advantages of additive manufacturing in blade production is its ability to create intricate and optimized designs [16, 22, 31]. Traditional blade manufacturing processes often involve the use of molds, which restrict design flexibility and limit the achievable geometries. In contrast, additive manufacturing allows for the production of complex, free-form shapes that can be tailored to enhance aerodynamic performance and energy capture. This capability enables the design of blades with improved efficiency and reduced aerodynamic drag, resulting in increased power output. Some global examples of additive manufacturing techniques being used in wind turbine blade production are the following:

- Siemens Gamesa Renewable Energy (SGRE): Siemens Gamesa, a leading wind turbine manufacturer, has been exploring additive manufacturing for blade production. They have partnered with companies like Oak Ridge National Laboratory in the United States to develop 3D printing techniques for manufacturing wind turbine blades. The goal is to optimize blade designs, reduce production costs, and enhance overall turbine performance.
- Vestas: Vestas, another prominent wind turbine manufacturer, has also ventured into additive manufacturing for blade production. They have collaborated with the Technical University of Denmark to develop 3D printing techniques that enable the production of complex blade geometries. This approach allows Vestas to optimize blade designs for specific wind conditions and improve energy capture.

- GE Renewable Energy: GE Renewable Energy has been exploring the use of additive manufacturing for wind turbine blades. They have partnered with Oak Ridge National Laboratory and Sandia National Laboratories in the United States to develop new manufacturing techniques. GE aims to leverage additive manufacturing to create larger and more efficient blades, leading to increased energy generation from wind turbines.
- ORE Catapult: The Offshore Renewable Energy (ORE) Catapult, based in the United Kingdom, has been actively involved in researching and testing additive manufacturing techniques for wind turbine blades [17]. They have conducted projects focused on the development of large-scale 3D printing methods for blade production. Their efforts aim to enhance blade performance, reduce costs, and accelerate the deployment of wind energy technologies.

Furthermore, additive manufacturing offers the potential for lightweight blade structures. By utilizing advanced lattice structures and optimizing material distribution, it is possible to reduce the weight of the blades without compromising their structural integrity. Lighter blades require less energy to rotate, allowing for improved turbine performance and efficiency. Additionally, the reduced weight of the blades can lead to lower material costs and transportation expenses. Another significant advantage of additive manufacturing is the potential for customization. With traditional manufacturing processes, blades are often produced in large quantities with standardized designs. However, additive manufacturing enables the production of unique blades tailored to specific site conditions and wind profiles. This customization can optimize energy capture and maximize the performance of wind turbines in varying environments. Additionally, customization can lead to a reduction in maintenance costs and downtime, as the blades can be designed to better withstand specific wind conditions and loads. Additive manufacturing also offers the opportunity to reduce material waste during the manufacturing process. Traditional blade production methods often result in significant material waste, as excess material is trimmed or discarded during the shaping and molding process. In contrast, additive manufacturing is an additive process, where materials are added layer by layer, resulting in minimal waste generation. This reduction in material waste contributes to the sustainability of blade production by minimizing resource consumption and environmental impact.

Furthermore, additive manufacturing can streamline the supply chain and reduce lead times in blade production. With traditional manufacturing processes, the production of molds and tooling can be time-consuming and expensive. Additive manufacturing eliminates the need for molds, allowing for more rapid prototyping and production. This increased efficiency can accelerate the development and deployment of wind turbine blades, contributing to the growth of the wind energy industry. While additive manufacturing holds great promise for blade production, there are still challenges to overcome [66, 67, 68–75, 76–78]. The scale-up of additive manufacturing for large wind turbine blades remains a technical and logistical challenge. Currently, the size limitations of additive manufacturing equipment restrict the production of full-scale blades. However, advancements in technology and the development of large-scale additive manufacturing systems are underway to address this limitation.

7.5 Challenges in Implementing Reusable Blade Technologies

7.5.1 Structural Integrity of Reusable Blades

Structural integrity is a critical aspect of reusable wind turbine blades as it directly impacts their performance, reliability, and safety. Wind turbine blades experience complex and dynamic loading conditions during operation, including aerodynamic forces, gravitational loads, and mechanical stresses. Therefore, it is essential to design and manufacture blades that can withstand these forces and maintain their structural integrity throughout their operational lifespan [18, 27, 79]. When implementing reusable blade technologies, one of the key challenges is to ensure that the new materials and construction techniques used meet the required strength and stiffness criteria. These materials may have different mechanical properties and behaviors compared to traditional blade materials. Therefore, a thorough evaluation of their structural performance is necessary to ensure that they can withstand the operational forces and maintain their shape and functionality. Advanced computer simulations, such as finite element analysis (FEA), play a crucial role in assessing the structural behavior of reusable blade designs [80, 81]. FEA models simulate the interaction between the blade structure and the applied loads, enabling engineers to evaluate stress distribution, deflections, and failure mechanisms. These simulations help optimize the blade design, identify potential weak points, and refine the structural elements to enhance performance and durability.

In addition to computer simulations, physical testing is performed to validate the structural integrity of reusable blade designs. Static load tests apply controlled forces to the blades to evaluate their stiffness, strength, and deformation under different loading scenarios. Dynamic load tests, on the other hand, subject the blades to cyclic loading conditions that simulate real-world wind conditions. These tests assess the fatigue behavior of the blades and their ability to withstand repeated loading without failure. Continuous monitoring and inspection techniques are employed throughout the operational life of the blades to detect any signs of structural degradation or damage. Structural health monitoring systems, such as strain gauges, accelerometers, and acoustic emission sensors, are used to monitor the behavior of the blades in real time. These systems can detect changes in structural characteristics, such as stress levels, deflections, and vibration patterns, which may indicate potential issues. Early detection allows for proactive maintenance actions, such as repairs or reinforcements, to be undertaken to prevent further damage and ensure the structural integrity of the blades.

Furthermore, the structural integrity of reusable blades is closely linked to their manufacturing processes. Consistency and quality control during manufacturing are crucial to achieving the desired structural performance. Manufacturing defects, such as voids, delamination, or inconsistencies in fiber orientation, can compromise the structural integrity of the blades. Therefore, strict quality control measures are implemented to ensure that the manufacturing processes meet the required standards and specifications. As the wind energy industry strives for larger and more efficient turbines, the structural integrity of the blades becomes even more critical. Longer blades are subjected to higher loads and increased bending moments, which can pose additional challenges in maintaining structural integrity. Advanced design techniques, such as load control strategies and aerodynamic profiling, are employed to optimize the blade shape and reduce stress concentrations in critical areas. Additionally, materials with enhanced mechanical properties are being developed to improve the structural integrity of reusable blades—for example, carbon fiber-reinforced polymers (CFRP) offer higher strength-to-weight ratios compared to traditional materials. The use of CFRP can enable the design of longer and more efficient blades that can capture more wind energy while maintaining structural integrity.

7.5.2 Fatigue Resistance and Durability

Fatigue resistance and durability are crucial factors in the successful implementation of reusable wind turbine blades. Wind turbines operate in harsh environmental conditions and are subjected to cyclic loading, which can lead to fatigue damage over time [20, 28, 82]. Ensuring that blades can withstand these conditions and maintain their structural integrity is essential for the long-term performance and reliability of wind farms. Fatigue damage occurs when a material undergoes repeated cyclic loading, causing progressive microscopic cracks to initiate and propagate. The cyclic loading experienced by wind turbine blades includes the fluctuating forces induced by wind gusts, turbulence, and gravitational loads. Over time, these cyclic loads can lead to the growth of cracks, which, if left undetected or unaddressed, can ultimately result in blade failure. To address the challenges associated with fatigue resistance and durability, several strategies and technologies are employed during the design, manufacturing, and operational phases of wind turbine blades.

In the design phase, engineers employ various techniques to enhance the fatigue resistance of blades. Load control strategies aim to reduce the variations in loads experienced by the blades. By optimizing the design, including the blade shape, twist, and aerodynamic profile, engineers can reduce the occurrence of high-stress regions and minimize the risk of fatigue failure. Additionally, the use of advanced materials, such as carbon fiber-reinforced polymers (CFRP), can improve the fatigue performance of blades due to their higher strength and stiffness compared to traditional materials. During the manufacturing process, quality control measures are implemented to ensure the integrity of the blades. This includes stringent inspection procedures to detect any defects or inconsistencies that could compromise the fatigue resistance of the blades. Ultrasonic testing, X-ray inspection, and other non-destructive testing methods are used to identify any internal defects, such as voids or delamination, that could serve as stress concentration points and initiate fatigue cracks. Furthermore, advancements in manufacturing techniques, such as automated processes and robotic manufacturing, contribute to improved blade durability. These techniques ensure consistent and high-quality blade production, reducing the likelihood of manufacturing defects that could affect the fatigue performance of the blades.

Once in operation, continuous monitoring and inspection methods are employed to assess the fatigue behavior and structural health of the blades. Structural health monitoring systems, including strain gauges, accelerometers, and vibration sensors, are installed on the blades to measure and analyze their response to operational loads. These systems provide real-time data on the structural performance of the blades, allowing operators to detect any signs of fatigue damage or performance degradation. Additionally, advanced monitoring techniques, such as acoustic emission testing and thermographic inspections, can be used to identify and locate potential fatigue cracks or defects within the blades. These techniques enable the early detection of damage, allowing for timely maintenance and repair actions to be taken before the integrity of the blades is compromised. To further enhance fatigue resistance and durability, ongoing research is focused on developing innovative materials and technologies—for example, self-healing materials are being explored to improve the fatigue performance of blades. These materials have the ability to autonomously repair minor damage, such as microcracks, by triggering a healing response when subjected to external stimuli.

Furthermore, advancements in predictive modeling and simulation techniques are being used to assess the long-term fatigue behavior of blades. By combining real-world operational data with numerical simulations, engineers can better understand the fatigue life of the blades and develop more accurate predictive models. This information allows for optimized maintenance and inspection schedules, reducing downtime and optimizing the overall performance of wind farms. The implementation of advanced materials, such as carbon nanotubes and graphene, is also being explored for their potential in enhancing fatigue resistance and durability. These nanomaterials possess exceptional mechanical properties and could provide increased strength, stiffness, and fatigue resistance compared to traditional blade materials. However, further research and development are needed to fully understand their behavior and ensure their reliable and cost-effective integration into blade designs.

7.5.3 Manufacturing Scalability and Cost-Effectiveness

Manufacturing scalability and cost-effectiveness are key considerations in the implementation of reusable wind turbine blades. To achieve the widespread adoption of these blades and maximize the benefits of renewable energy, it is essential to develop manufacturing processes that can efficiently produce high-quality blades at a reasonable cost. The manufacturing of wind turbine blades involves several complex and labor-intensive processes. These include mold preparation, material layup, resin infusion, curing, finishing, and quality control [30, 34, 36]. Traditional manufacturing methods, such as hand layup and vacuum infusion, have been commonly used in blade production. However, these methods can be time-consuming, require skilled labor, and may result in variations in blade quality. To overcome these challenges and improve manufacturing scalability, advanced technologies and innovative approaches are being explored in the production of reusable wind turbine blades. One such technology is automated manufacturing, which involves the use of robotic systems and advanced machinery to streamline the production process. Robotic systems can precisely and consistently perform tasks such as material layup, resin infusion, and trimming, resulting in higher manufacturing accuracy and reduced labor costs. Automation also enables increased production rates, allowing for the scalability of blade manufacturing to meet the growing demand for wind energy.

Furthermore, additive manufacturing, also known as 3D printing, holds significant potential in revolutionizing blade production. This technology enables the fabrication of complex geometries and customized designs with reduced material waste. Additive manufacturing allows for the direct production of blade components, eliminating the need for molds and reducing tooling costs. It also offers flexibility to iterate and optimize designs quickly, leading to improved blade performance and cost-effectiveness. Several research and development efforts are underway to explore the feasibility of additive manufacturing for wind turbine blades. Advanced materials, such as carbon fiber-reinforced thermoplastics, are being investigated for their suitability in additive manufacturing processes. These materials offer enhanced mechanical properties, including improved fatigue resistance, and can be processed using additive manufacturing techniques. Cost-effectiveness is a significant consideration in the manufacturing of reusable blades, as it directly impacts the economic viability of wind energy projects. The goal is to produce blades that are not only environmentally friendly but also financially competitive with conventional blade technologies. One aspect of cost-effectiveness is the selection of materials. Advanced composite materials, such as carbon fiber-reinforced polymers (CFRP), have desirable properties for blade applications, including high strength-to-weight ratios and fatigue resistance. While these materials may have higher upfront costs compared to traditional materials, their longer lifespan and improved performance can contribute to the overall cost-effectiveness of wind turbine blades. Furthermore, advancements in material manufacturing and supply chain management can help reduce material costs over time.

Another aspect is the optimization of manufacturing processes to minimize production time and costs. Through the integration of automation, advanced machinery, and streamlined workflows, manufacturers can improve production efficiency and reduce labor costs. The use of digital tools, such as computer-aided design and simulation software, can also aid in optimizing blade designs and manufacturing processes, leading to cost savings. Furthermore, the development of standardized manufacturing processes and modular blade designs can contribute to cost-effectiveness.

Standardization allows for efficient production and reduces the need for customization, resulting in economies of scale. Modular designs enable the assembly of blades from standardized components, simplifying manufacturing and maintenance processes. These approaches can help lower production costs and enhance the scalability of blade manufacturing. In addition to manufacturing scalability and cost-effectiveness, considerations for the overall life cycle costs of wind turbine blades are also important. This includes the costs associated with transportation, installation, operation, maintenance, and end-of-life management. By optimizing the design and manufacturing processes, manufacturers can reduce these life cycle costs and improve the economic viability of wind energy projects.

7.6 Implications of Reusable Wind Turbine Blades

7.6.1 Cost Reduction and Enhanced Energy Production

The implementation of reusable wind turbine blades brings several implications that can lead to cost reduction and enhanced energy production in the wind energy sector [83–85]. These implications include improved efficiency, increased reliability, reduced maintenance costs, and extended operational life. Reusable blade technologies aim to optimize the design and manufacturing processes, resulting in more efficient blades. Enhanced aerodynamic profiles, longer and lighter blades, and improved material properties contribute to increased energy capture from the wind. With higher energy production, wind farms can generate more electricity, which translates into greater revenue potential. Moreover, reusable blades often incorporate advanced materials, such as carbon fiber-reinforced polymers (CFRP), that offer superior mechanical properties compared to traditional materials. These materials exhibit higher strength-to-weight ratios, enabling the construction of longer and more efficient blades. The increased length of the blades allows for a larger swept area, meaning they can capture more wind energy, leading to enhanced energy production. Furthermore, the improved reliability of reusable blades can result in reduced maintenance costs. By utilizing advanced materials and manufacturing processes, the blades can withstand the dynamic forces exerted by the wind and maintain their structural integrity for longer periods. This minimizes the need for frequent inspections, repairs, and replacements, resulting in decreased downtime and associated maintenance expenses. The extended operational life of reusable blades also contributes to cost reduction by deferring the need for blade replacement.

Additionally, advancements in manufacturing scalability and cost-effectiveness can drive down the overall cost of wind turbine blades. Automation, additive manufacturing, and standardization of processes allow for increased production rates, reduced labor costs, and optimized material usage. As a result, the upfront costs associated with reusable blade technologies can be lowered, making wind energy projects more economically viable. The cost reduction implications of reusable wind turbine blades contribute to the competitiveness of wind energy in the global market. As the levelized cost of electricity (LCOE) decreases, wind power becomes more attractive compared to conventional energy sources [33, 46]. This has a significant impact on energy transition efforts by providing affordable and sustainable electricity generation solutions. One of the key factors contributing to cost reduction is the increased efficiency of reusable blades. Improved aerodynamic designs and longer blade lengths allow for higher energy capture from the wind. This means that wind farms equipped with reusable blades can produce more electricity per turbine, resulting in increased revenue potential. Additionally, advancements in blade manufacturing processes, such as precision molding and additive manufacturing techniques, enable the production of blades with optimized shapes and profiles, further enhancing their energy capture capabilities. The use of advanced materials, such as carbon fiber-reinforced polymers (CFRP), in reusable blades also contributes to cost reduction. CFRP materials offer superior strength-to-weight ratios compared to traditional materials like fiberglass composites. This allows for the construction of longer and lighter blades that can withstand the dynamic forces exerted by the wind. The reduced weight of the blades not only improves their aerodynamic performance but also reduces transportation costs during manufacturing and installation.

Furthermore, the improved reliability and extended operational life of reusable blades contribute to cost reduction in the long run. By utilizing advanced materials and manufacturing techniques, reusable blades can withstand the stresses and loads experienced during operation, minimizing the need for frequent repairs and replacements. This reduces downtime and associated maintenance costs, improving the overall profitability of wind farms. Advancements in manufacturing scalability and cost-effectiveness play a significant role in reducing the upfront costs of reusable blades [46, 54]. Automation and standardized manufacturing processes enable higher production rates, reducing labor costs and increasing efficiency. Additive manufacturing techniques, such as 3D printing, offer opportunities for more precise and cost-effective blade production. These technologies allow for the optimization of material usage, waste reduction, and improved quality control. Additionally, standardized manufacturing processes enable economies of scale, further driving down production costs. The cost reduction implications of reusable blades have significant implications for the wind energy industry [37]. Lowering the levelized cost of electricity (LCOE) makes wind energy more competitive in the global energy market. This can attract more investments in wind power projects and accelerate the transition to renewable energy sources. The increased cost-effectiveness of wind energy also contributes to energy security and economic development by reducing reliance on fossil fuels and creating job opportunities in the renewable energy sector.

7.6.2 Environmental Benefits and Reduction of Carbon Emissions

The implementation of reusable wind turbine blades brings significant environmental benefits and contributes to the reduction of carbon emissions. By harnessing the power of wind, renewable energy technologies like wind turbines offer a clean and sustainable alternative to conventional fossil fuel-based power generation. Reusable blades further enhance these environmental benefits through improved efficiency, reduced material waste, and lower carbon footprint. One of the primary environmental benefits of wind energy is the reduction of carbon emissions. Wind power is a carbon-neutral energy source, which means that it does not release greenhouse gases (GHGs) during electricity generation. Unlike fossil fuel-based power plants, wind turbines do not burn fuel and therefore do not produce carbon dioxide (CO_2) or other pollutants that contribute to climate change and air pollution. The use of reusable blades in wind turbines helps to maximize energy production and further reduce the need for fossil fuel-based electricity generation, leading to a significant reduction in carbon emissions. By increasing the energy production efficiency of wind turbines, reusable blades contribute to a greater displacement of fossil fuel-based power generation. As more wind energy is integrated into the grid, the demand for electricity generated from fossil fuels decreases. This results in a direct reduction in CO_2 emissions, which is crucial for mitigating climate change and achieving global emission reduction targets.

Reusable blades also have indirect environmental benefits through the reduction of material waste. The manufacturing and disposal of wind turbine blades can generate significant amounts of waste, which can have environmental implications if not properly managed. Traditional fiberglass blades, in particular, present challenges in terms of recyclability and end-of-life disposal. However, reusable blades often incorporate advanced materials and manufacturing processes that aim to address these issues—for example, the use of carbon fiber-reinforced polymers (CFRP) in reusable blades offer superior mechanical properties and improved recyclability compared to fiberglass composites [26, 35, 45, 52]. The carbon fibers used in CFRP blades can be recovered and reused in other applications, reducing waste and the environmental impact associated with blade disposal. This helps to minimize the accumulation of blade waste and promotes a more sustainable approach to wind turbine technology. Furthermore, advancements in recycling technologies for composite materials are being explored to increase the recyclability of wind turbine blades. Various methods, such as mechanical shredding, chemical processes, and pyrolysis, are being developed to effectively separate and recover the valuable components of composite materials. These recycling efforts contribute to the circular economy by reducing the reliance on virgin materials and minimizing the environmental footprint of wind turbine blades. In addition to carbon emission reduction and waste management, wind energy with reusable blades offers other environmental benefits. Wind turbines have a smaller land footprint compared to traditional power plants, making them a favorable option for energy production in terms of land use. The land around wind turbines can still be used for agriculture, grazing, or other purposes, allowing for dual land use and minimizing the impact on ecosystems.

Moreover, wind energy does not consume water resources for power generation, unlike fossil fuel-based plants that require large amounts of water for cooling and other processes. This is particularly important in regions facing water scarcity or where water resources need to be conserved for other purposes such as agriculture or human consumption. By utilizing wind energy with reusable blades, water resources can be preserved, contributing to sustainable water management. The environmental benefits of reusable wind turbine blades extend beyond the direct impacts of carbon emission reduction and waste management. They promote the development and deployment of renewable energy technologies, helping to diversify the energy mix and reduce dependence on fossil fuels. This leads to a more sustainable and resilient energy system that is less vulnerable to price fluctuations and supply disruptions associated with fossil fuels. Furthermore, the environmental benefits of wind energy contribute to global efforts to mitigate climate change and achieve sustainable development goals. The reduction of carbon emissions helps to limit the increase in global temperatures, minimize the frequency and intensity of extreme weather events, and protect ecosystems and biodiversity. Wind energy with reusable blades aligns with international agreements and commitments to combat climate change, such as the Paris Agreement and the United Nations Sustainable Development Goals.

7.6.3 Policy Frameworks and Industry Collaboration

Policy frameworks and industry collaboration play a vital role in promoting the implementation of reusable wind turbine blades and driving the transition to a more sustainable energy future. Governments, regulatory bodies, and industry stakeholders need to work together to create supportive policies, regulations, and incentives that encourage the adoption of reusable blade technologies, facilitate research and development, and promote collaboration across the wind energy sector. At the national level, governments can establish renewable energy targets and incentivize the deployment of wind energy through mechanisms such as feed-in tariffs, renewable portfolio standards, and tax incentives. These policies provide a favorable market environment for wind energy projects, including those utilizing reusable blades. Governments can also support research and development initiatives focused on advancing blade technologies, materials, and manufacturing processes. Funding programs, grants, and partnerships between public and private entities can facilitate innovation and accelerate the commercialization of reusable blades.

International collaboration and knowledge sharing are crucial in driving the development and adoption of reusable blade technologies. Governments and industry associations can facilitate collaboration through platforms such as international conferences, workshops, and research networks. These platforms provide opportunities for stakeholders from different countries to exchange ideas, share best practices, and collaborate on research and development projects. By pooling resources and expertise, the global wind energy community can collectively address common challenges and accelerate the implementation of reusable blade technologies. Furthermore, the wind energy industry itself plays a significant role in driving the adoption of reusable blades through collaboration and knowledge sharing. Wind turbine manufacturers, developers, and operators can work together to establish industry standards and best practices for the design, manufacturing, and deployment of reusable blades. Sharing data and experiences related to the performance, durability, and cost-effectiveness of reusable blade technologies can contribute to a collective understanding of their benefits and challenges. Industry associations and research institutions can also play a crucial role in facilitating collaboration and knowledge sharing. They can establish platforms for information exchange, provide technical support and guidance, and conduct research and testing to validate the performance and reliability of reusable blades. Collaborative research projects, joint ventures, and public-private partnerships can drive innovation and accelerate the adoption of reusable blade technologies.

In addition to policy frameworks and industry collaboration, public awareness and education are essential to garner support for reusable blade technologies. Governments, industry associations, and environmental organizations can conduct public outreach campaigns to raise awareness about the environmental benefits of wind energy and the role of reusable blades in reducing carbon emissions. Public support and understanding can influence policymakers, create market demand, and drive investments in wind energy projects. Moreover, integrating wind energy into the broader energy system requires collaboration and coordination with other stakeholders, including grid operators, utilities, and energy storage providers. Reusable blade technologies can contribute to grid stability and flexibility by enhancing the efficiency and reliability of wind turbines. Collaborative efforts are needed to optimize the integration of wind energy into the grid, develop storage solutions, and implement smart grid technologies that enable the effective management of renewable energy resources. Policy frameworks and industry collaboration should also address the challenges associated with the deployment of reusable blade technologies. These challenges include the initial cost of implementing new technologies, the need for standardized recycling processes, and the development of a skilled workforce and supply chains. Governments can provide financial support and incentives to help overcome the initial cost barriers, promote research and development of recycling technologies, and invest in training programs to develop a skilled workforce.

7.7 Testing, Modeling, and Simulation for Reliable Reusable Blade Designs

7.7.1 Importance of Rigorous Testing

Rigorous testing is crucial in the development of reliable reusable blade designs for wind turbines due to the demanding operational conditions and loads that the blades experience. Wind turbine blades must be able to withstand a wide range of forces, including dynamic wind forces, turbulent flows, and gravitational forces while maintaining their structural integrity, performance, and durability. Thorough testing throughout the design and development process is essential to identify and address any design flaws, weaknesses, or areas in need of improvement. The testing process begins during the design phase, where prototypes of the reusable blades are manufactured and subjected to various laboratory and field tests. These tests serve multiple purposes, including evaluating the mechanical properties, assessing the structural behavior, and examining the performance characteristics of the blades. Static load testing is a common method used to evaluate the structural strength of wind turbine blades. In this type of testing, the blade is subjected to predetermined static loads that simulate the effects of wind forces and gravitational loads. The response of the blade is carefully monitored, and structural deformations and stresses are measured. Static load testing helps validate the blade's ability to withstand extreme loads and ensures its structural integrity. By applying carefully calibrated loads, engineers can determine the blade's load-bearing capacity and identify any potential failure points.

Dynamic load testing is another critical aspect of blade testing. Wind turbine blades experience cyclic loading conditions due to the dynamic nature of wind forces. During dynamic load testing, the blade is subjected to a range of load cases that simulate various operating conditions, including gusts, turbulent flows, and fatigue loading. This type of testing is crucial for assessing the fatigue resistance and durability of the blades over their intended operational lifespan. Monitoring systems are used to measure the response of the blade to dynamic loading, including vibrations and deflections. By subjecting the blade to realistic loading scenarios, engineers can evaluate its performance under different conditions and identify any weaknesses or potential points of failure. Field testing is an essential component of blade development. Full-scale blades are installed on operational wind turbines in real-world wind farm environments, and their performance, power output, and structural behavior are closely monitored. Field testing provides valuable data that cannot be fully replicated in laboratory conditions, allowing engineers to evaluate the blade's behavior in actual operating environments [40, 50]. This data can be used to validate and refine the design of reusable blades, ensuring that their performance matches expectations and requirements. It is important to note that testing protocols and standards for reusable blades may differ from those used for conventional blades. This is due to the use of new materials, construction techniques, and design features in reusable blade technologies. Therefore, it is necessary to adapt testing protocols and standards to account for these differences and ensure that the blades meet the necessary safety, performance, and reliability requirements. The testing process should be carefully designed and executed to address the unique characteristics of reusable blades and provide accurate and reliable results.

7.7.2 Modeling and Simulation Techniques for Design Optimization

Modeling and simulation techniques play a crucial role in the design optimization of wind turbine blades. They provide valuable insights into the structural behavior, performance, and aerodynamic characteristics of the blades, enabling engineers to refine and optimize their design. By using advanced computational tools, engineers can simulate various operating conditions and evaluate the response of the blades to different loads, improving their efficiency, reliability, and overall performance. One of the primary goals of modeling and simulation is to assess the structural integrity of the blades. Finite element analysis (FEA) is a widely used technique for modeling and simulating the structural behavior of wind turbine blades [38, 51]. FEA divides the blade into smaller elements, allowing engineers to analyze the stresses, strains, and deformations within the structure. By applying appropriate boundary conditions and load cases, FEA can accurately predict the response of the blade under different operating conditions. This enables engineers to identify potential failure points and make design modifications to improve the blade's structural integrity. Aerodynamic performance is another critical aspect of wind turbine blade design optimization. Computational fluid dynamics (CFD) is a powerful tool that simulates the flow of air around the blade and predicts its aerodynamic performance. By solving the governing equations for fluid flow, CFD can calculate important parameters such as lift, drag, and flow separation. This information helps engineers optimize the blade shape, airfoil profiles, and other design features to improve energy capture and reduce aerodynamic losses. CFD simulations can also provide insights into the impact of turbulent flow, yaw angles, and other factors on the blade's performance.

In addition to structural and aerodynamic simulations, modeling and simulation techniques can also be used to assess the dynamic behavior of wind turbine blades. Dynamic simulations allow engineers to study the response of the blades to external forces, including wind gusts and turbulent flows. These simulations help evaluate the fatigue resistance, natural frequencies, and vibration characteristics of the blades. By understanding how the blades behave dynamically, engineers can optimize their design to reduce fatigue loads and enhance their durability. Furthermore, optimization algorithms can be employed in conjunction with modeling and simulation techniques to automatically search for the best design parameters and configurations. These algorithms explore a range of design variables, such as blade length, airfoil profiles, twist distribution, and structural reinforcements, to find the optimal combination that maximizes energy capture and minimizes structural loads. Optimization techniques, such as genetic algorithms or gradient-based methods, can efficiently search the design space and provide valuable insights for design improvements. The use of modeling and simulation techniques for design optimization offers several advantages. First, it enables engineers to evaluate a wide range of design options without the need for physical prototypes, reducing the time and cost associated with experimental testing. Virtual simulations provide a faster and more efficient means of exploring different design configurations and making informed design decisions.

Second, simulations allow engineers to study the performance of the blades under various operating conditions, including extreme events that are challenging to replicate in physical testing. This helps ensure that the blades can withstand and perform reliably in real-world scenarios. Third, modeling and simulation techniques provide a deeper understanding of the underlying physics and behavior of wind turbine blades. They allow engineers to visualize and analyze the flow patterns, stress distributions, and structural responses, providing valuable insights into the performance limitations and potential areas for improvement. It is important to note that the accuracy of modeling and simulation results depends on the quality of the input data, including material properties, boundary conditions, and environmental factors. Therefore, it is crucial to validate the simulation models and compare their predictions with experimental data to ensure their reliability and accuracy. Experimental data can be obtained from laboratory tests, field measurements, or reference studies to validate the simulation results and improve confidence in the design optimization process.

7.8 Future Prospects and Research Directions

7.8.1 Interdisciplinary Approaches for Sustainable Innovation

Interdisciplinary approaches are crucial for achieving sustainable innovation in wind turbine blade design. The challenges associated with developing reusable wind turbine blades require the collaboration and integration of various disciplines, including materials science, mechanical engineering, aerodynamics, environmental science, and renewable energy systems. By bringing together experts from these diverse fields, interdisciplinary approaches can lead to the development of holistic solutions that enhance the performance, recyclability, and sustainability of wind turbine blades. One of the key areas where interdisciplinary approaches are essential is materials science and engineering. Materials scientists play a crucial role in developing advanced composite materials with superior mechanical properties, recyclability, and reduced environmental impact. They explore new materials, such as bio-based composites, nanomaterials, and recyclable polymers, to improve the overall performance and sustainability of wind turbine blades. Collaborating with mechanical engineers, materials scientists can optimize the material selection, manufacturing processes, and structural designs to ensure that the blades are strong, lightweight, and durable.

Aerodynamics is another important field that benefits from interdisciplinary collaboration. Aerodynamicists study and optimize the airflow around wind turbine blades to maximize energy capture and minimize aerodynamic losses. Computational fluid dynamics (CFD) simulations are used to model and analyze the complex airflow, allowing engineers to refine the blade shape, airfoil profiles, and other design features. Collaboration between aerodynamicists, materials scientists, and mechanical engineers enables the creation of blade designs that balance aerodynamic efficiency, structural integrity, and sustainability. Moreover, environmental experts contribute to interdisciplinary approaches by assessing the environmental impact of wind turbine blades throughout their lifecycle. Life cycle assessments (LCAs) can be conducted to evaluate the overall environmental footprint, considering factors such as energy consumption, carbon emissions, waste generation, and resource depletion. By integrating LCA findings into the design process, engineers can identify opportunities for minimizing the environmental impact and improving the sustainability of wind turbine blades. Collaboration between environmental experts, materials scientists, and mechanical engineers ensures that the blades are designed with a comprehensive understanding of their environmental implications.

Renewable energy systems and grid integration expertise are also critical in interdisciplinary approaches to wind turbine blade design. Collaboration between wind turbine manufacturers, power grid operators, and energy system experts can facilitate the development of blade designs that are compatible with grid requirements and contribute to the stability and reliability of the overall power system. By optimizing the interaction between wind turbines and the grid, the performance and efficiency of wind turbine blades can be enhanced. The integration of renewable energy systems and grid integration expertise into interdisciplinary teams ensures that blade designs align with the broader goals of the renewable energy industry. Interdisciplinary approaches foster innovation by encouraging the exchange of knowledge, expertise, and resources. Research institutions, universities, and industry partners can collaborate on joint research projects, sharing data, findings, and resources to develop new materials, manufacturing techniques, and design approaches [86, 87]. By combining their expertise and resources, interdisciplinary teams can address complex challenges, explore new ideas, and push the boundaries of wind turbine blade technology. Engineers from wind turbine manufacturers, blade design firms, and renewable energy companies play a crucial role in interdisciplinary collaborations. Their practical experience in manufacturing, installation, and operation of wind turbines provides valuable insights and industry perspectives. By working alongside researchers, engineers contribute their real-world expertise to the development of blade designs that are not only technologically advanced but also practical and cost-effective. Their involvement ensures that the proposed solutions align with industry requirements and can be implemented on a larger scale. In addition to researchers and engineers, engaging stakeholders is vital in shaping the future of reusable wind turbine blades. Stakeholders include policymakers, regulatory bodies, environmental organizations, local communities, and end-users. By involving stakeholders in the decision-making process, their perspectives and concerns can be addressed, ensuring that blade designs meet societal needs and are socially and environmentally responsible. Collaboration with stakeholders promotes transparency, fosters public acceptance, and facilitates the adoption of sustainable wind turbine blade designs.

7.8.2 Collaboration Among Researchers, Engineers, and Stakeholders

Collaboration among researchers, engineers, and stakeholders is essential for the successful development and implementation of sustainable wind turbine blade technologies. The complexity of the challenges involved requires the collective efforts, expertise, and perspectives of various stakeholders to drive innovation, address technical barriers, and ensure the long-term viability of reusable blade designs. Researchers play a crucial role in advancing the scientific understanding and technological capabilities of wind turbine blades [41, 88]. They conduct fundamental research, explore new materials and manufacturing techniques, and develop modeling and simulation tools to optimize blade performance. Collaboration among researchers from different institutions and disciplines allows for the exchange of knowledge, sharing of resources, and validation of research findings. By working together, researchers can push the boundaries of knowledge and drive advancements in wind turbine blade technology. Engineers, on the other hand, bring practical expertise and industry experience to the collaboration. They understand the operational challenges, manufacturing processes, and cost considerations associated with wind turbine blades. Engineers from wind turbine manufacturers, blade design firms, and renewable energy companies contribute valuable insights into the feasibility and scalability of new technologies. Their involvement ensures that research outcomes align with industry requirements, standards, and practical constraints. Collaborating with engineers provides researchers with real-world perspectives, enabling them to develop blade designs that are not only technologically advanced but also practical, cost-effective, and manufacturable.

Stakeholders, including policymakers, regulatory bodies, environmental organizations, local communities, and end-users, also play a crucial role in the collaboration. Their engagement is essential for shaping the direction and impact of wind turbine blade technologies. Policymakers and regulatory bodies create the policy frameworks and standards that govern the deployment and operation of wind turbines. Engaging with them allows researchers and engineers to understand the regulatory landscape and ensure compliance with relevant regulations. Environmental organizations provide valuable insights into the ecological and social implications of wind turbine blades, helping to address concerns and develop sustainable solutions. Local communities and end-users offer perspectives on the social acceptance, environmental impact, and economic benefits of wind turbine installations. Collaborating with stakeholders ensures that the blade designs reflect societal needs, address concerns, and maximize the benefits for all involved parties. Collaboration among researchers, engineers, and stakeholders can take various forms, including joint research projects, industry–academia partnerships, public–private collaborations, and advisory committees. These collaborative efforts facilitate the exchange of knowledge, data, and resources. Researchers can gain access to industry expertise, real-world data, and test facilities, enabling them to validate their research findings and enhance the practical applicability of their work. Engineers can benefit from the latest research advancements and innovative ideas, allowing them to improve the design, performance, and reliability of wind turbine blades. Stakeholders can contribute their perspectives, provide valuable feedback, and influence the direction of research and development activities. Here are a few global examples of collaboration among researchers, engineers, and stakeholders in the field of wind turbine blade technologies:

- European Wind Energy Association (EWEA): The EWEA brings together researchers, industry representatives, policymakers, and other stakeholders to promote the development and deployment of wind energy in Europe. They facilitate collaboration through working groups, conferences, and research projects focused on various aspects of wind turbine technology, including blade design and optimization.
- Sandia National Laboratories and National Renewable Energy Laboratory (NREL)—United States: These institutions collaborate on numerous research projects related to wind energy, including wind turbine blade design, materials testing, and structural analysis. Their collaboration involves the sharing of research findings, joint testing facilities, and cooperative research programs.
- Offshore Renewable Energy (ORE) Catapult—United Kingdom: ORE Catapult is a leading research and innovation center focused on advancing offshore renewable energy technologies. They collaborate with academic institutions, industry partners, and government agencies to develop and test innovative wind turbine blade designs for offshore applications. Their collaborative approach helps accelerate technology development and de-risk new blade designs.
- China Wind Power Engineering Institute (CWPEI): CWPEI works closely with wind turbine manufacturers, research institutes, and industry associations in China to advance wind turbine technologies. They conduct joint research projects, share technical expertise, and collaborate on testing and certification programs. This collaboration helps drive technological advancements in blade design, structural analysis, and performance optimization.
- International Energy Agency (IEA)—Wind Task 37: Task 37 of the IEA's Wind Technology Collaboration Program focuses on wind energy systems engineering, including wind turbine design, operation, and maintenance. It brings together experts from different countries to collaborate on research, share best practices, and develop guidelines for wind turbine blade design and performance assessment.
- Wind Energy Institute of Canada (WEICan): WEICan collaborates with industry partners, research organizations, and government agencies to conduct research and testing on wind turbine technologies. Their collaboration includes

projects related to blade design optimization, materials testing, and performance evaluation. They provide a platform for industry–academia collaboration and contribute to the development of innovative and sustainable wind turbine blade technologies.

Furthermore, collaboration enables the alignment of objectives and the pooling of resources. Research institutions, universities, and industry partners can jointly fund research projects, leverage their respective capabilities, and share the costs associated with technology development. This collaborative approach helps to overcome financial barriers, accelerate research and development activities, and achieve breakthroughs that would be challenging to attain individually. By combining resources, stakeholders can collectively drive innovation, promote knowledge transfer, and ensure the successful implementation of sustainable wind turbine blade technologies. Collaboration also fosters innovation and promotes creativity [48, 89, 90]. By bringing together diverse perspectives and expertise, interdisciplinary teams can tackle complex problems from multiple angles. This collaborative environment encourages the exchange of ideas, sparks new insights, and pushes the boundaries of knowledge and technology. Researchers, engineers, and stakeholders can engage in brainstorming sessions, workshops, and conferences to explore innovative approaches, identify emerging trends, and address technical and societal challenges. The collaborative process encourages continuous learning, stimulates creativity, and leads to novel solutions that have a transformative impact on wind turbine blade design.

7.8.3 Potential Directions for Future Research

Future research in the field of wind turbine blade technologies holds great promise for further improving the performance, sustainability, and cost-effectiveness of renewable energy generation. Here are some potential directions for future research:

- Advanced materials development: Continued research into advanced materials is crucial for enhancing the performance and sustainability of wind turbine blades. The development of novel composite materials with improved mechanical properties, such as higher strength-to-weight ratios and enhanced fatigue resistance, can lead to longer and more efficient blade designs. Research efforts can focus on exploring alternative reinforcement fibers, such as carbon nanotubes, graphene, or natural fibers, to enhance the mechanical properties of the blades.

- Structural optimization and design: Further research is needed to optimize the structural design of wind turbine blades. This includes investigating new blade geometries, such as swept-tip or variable twist designs, to enhance aerodynamic efficiency and reduce loads. Advanced computational modeling techniques, such as machine learning and artificial intelligence, can be employed to optimize the shape, size, and distribution of materials within the blade structure. This can result in lighter, stronger, and more efficient blade designs.
- Aeroelasticity and dynamic response: Understanding the aeroelastic behavior and dynamic response of wind turbine blades is crucial for improving their performance and reliability. Future research can focus on developing more accurate and robust models to predict the blade's behavior under varying wind conditions, including turbulence and extreme events. This can help optimize blade designs to mitigate fatigue damage, reduce loads, and improve overall structural performance.
- Blade manufacturing and process innovation: Research efforts can be directed towards developing innovative manufacturing processes for wind turbine blades. Additive manufacturing techniques, such as 3D printing, hold the potential for more flexible and cost-effective production methods. Further research is needed to optimize additive manufacturing processes, improve material compatibility, and ensure the structural integrity of printed blades. Additionally, exploring automation and robotic technologies can help streamline blade manufacturing processes, reduce costs, and enhance quality control.
- Condition monitoring and maintenance strategies: Research can focus on developing advanced condition monitoring systems and maintenance strategies for wind turbine blades. This includes the integration of sensors, data analytics, and machine learning algorithms to detect and assess the structural health of the blades in real-time. By continuously monitoring the condition of the blades, maintenance activities can be optimized, and potential failures can be predicted

and prevented. This can result in improved operational reliability, reduced downtime, and lower maintenance costs.

- Offshore wind blade technologies: As the offshore wind industry continues to grow, research in offshore wind blade technologies becomes increasingly important. Future research can focus on developing blade designs that can withstand harsh marine environments, including saltwater corrosion, high wind speeds, and large waves. Additionally, exploring floating wind turbine technologies and adaptive blade designs for offshore applications can open new opportunities for harnessing wind energy in deeper waters.
- Sustainability and end-of-life solutions: As wind turbine installations increase worldwide, the issue of blade waste management becomes more critical. Research efforts can be directed toward developing sustainable end-of-life solutions for wind turbine blades, including recycling and repurposing strategies. Investigating the use of alternative materials with better recyclability and biodegradability can help reduce environmental impact. Additionally, research can focus on improving blade decommissioning techniques to minimize waste generation and facilitate efficient blade disposal or reuse.
- Integration with energy storage and grid systems: The integration of wind turbines with energy storage and grid systems is a key area for future research. Investigating the optimal integration of wind turbine blades with energy storage technologies, such as batteries or hydrogen storage, can enhance grid stability and enable more reliable and efficient energy generation. Research efforts can also explore the development of smart grid technologies and advanced control systems to enable better coordination and management of wind energy resources.

7.9 Conclusion

In conclusion, the development and implementation of reusable wind turbine blades hold immense potential for advancing the efficiency, sustainability, and cost-effectiveness of wind energy generation. The research and technological advancements in materials engineering, manufacturing techniques, and design optimization have paved the way for innovative and durable blade solutions. The implications of reusable wind turbine blades are far-reaching. They offer significant benefits in terms of cost reduction, enhanced energy production, environmental sustainability, and reduced carbon emissions. By improving the structural integrity, fatigue resistance, and recyclability of blades, the renewable energy industry can achieve higher energy yields, lower maintenance costs, and decreased reliance on fossil fuels.

However, several challenges need to be addressed to fully realize the potential of reusable blade technologies. The structural integrity of blades must be rigorously tested and validated through laboratory, field, and full-scale testing. This ensures that the blades can withstand the dynamic forces and loads they encounter during operation. The development and implementation of modeling and simulation techniques play a crucial role in optimizing blade design and performance. Collaboration among researchers, engineers, and stakeholders is essential for driving innovation and solving complex problems in the wind energy sector. By working together, sharing knowledge and resources, and fostering interdisciplinary approaches, the industry can accelerate progress and overcome technical and operational challenges. Examples of successful collaborations and partnerships at the national and international levels demonstrate the value of collective efforts in advancing wind turbine blade technologies. Future research directions should focus on interdisciplinary approaches, collaboration, and exploring new frontiers. Advanced materials development, structural optimization, aeroelasticity, manufacturing scalability, condition monitoring, and sustainability solutions are key areas that warrant further investigation. The integration of wind turbines with energy storage and grid systems also presents exciting opportunities for enhancing the reliability and efficiency of renewable energy generation. Moreover, policy frameworks and regulatory support play a vital role in fostering innovation and facilitating the adoption of reusable blade technologies. Governments, industry associations, and international organizations need to establish clear guidelines, incentives, and standards to promote sustainable practices, encourage investment, and drive the transition to renewable energy sources. Policy frameworks that prioritize research and development, incentivize collaboration and ensure environmental sustainability are crucial for the long-term success of wind energy projects. The continuous research, development, and collaboration efforts will drive the evolution of blade technologies and propel the wind energy industry toward a more sustainable and resilient future. By addressing the challenges, leveraging interdisciplinary approaches, and promoting collaboration among researchers, engineers, and stakeholders, we can unlock the full potential of reusable blade technologies and contribute to a cleaner, greener, and more sustainable energy landscape.

References

1. Prasad, B.B., Duvigneau, F., Juhre, D., Woschke, E., Application of particle dampers on a scaled wind turbine generator to improve low-frequency vibro-acoustic behavior. *Appl. Sci.*, 12, 671, 2022.
2. Karavida, S., Peponi, A., Wind turbine blade waste circularity coupled with urban regeneration: A conceptual framework. *Energies*, 16, 1–17, 2023.
3. Bledzki, A.K., Seidlitz, H., Krenz, J., Goracy, K., Urbaniak, M., Rösch, J.J., Recycling of carbon fiber reinforced composite polymers—review—part 2: Recovery and application of recycled carbon fibers. *Polymers (Basel)*, 12, 1–10, 2020.
4. Zhao, R., Zhang, H., Zhang, M.Y., Qu, F., Xu, Y., Competitor-weighted centrality and small-world clusters in competition networks on firms' innovation ambidexterity: Evidence from the wind energy industry. *Int. J. Environ. Res. Public Health.*, 20, 3339, 2023.
5. Bank, L.C., Arias, F.R., Yazdanbakhsh, A., Gentry, T.R., Al-Haddad, T., Chen, J.F., Morrow, R., Concepts for reusing composite materials from decommissioned wind turbine blades in affordable housing. *Recycling.*, 3, 3, 2018.
6. Psomopoulos, C.S., Kalkanis, K., Kaminaris, S., Ioannidis, G.C., Pachos, P., A review of the potential for the recovery of wind turbine blade waste materials. *Recycling.*, 4, 1–9, 2019.
7. De Fazio, D., Boccarusso, L., Formisano, A., Viscusi, A., Durante, M., A review on the recycling technologies of fibre-reinforced plastic (FRP) materials used in industrial fields. *J. Mar. Sci. Eng.*, 11, 851, 2023.
8. Li, X., Li, X., Mu, Z., Li, Y., Feng, F., An Experimental Study On Biochar/Polypyrrole Coating For Blade Anti-Icing Of Wind Turbines. *Coatings.*, 13, 759, 2023.
9. Pinto, T.H.L., Gul, W., Torres, L.A.G., Cimini, C.A., Ha, S.K., Experimental and numerical comparison of impact behavior between thermoplastic and thermoset composite for wind turbine blades. *Materials (Basel).*, 14, 6377, 2021.
10. Okokpujie, I.P., Tartibu, L.K., Aluminum Alloy Reinforced with Agro-Waste, and Eggshell as Viable Material for Wind Turbine Blade to Annex Potential Wind Energy: A Review. *J. Compos. Sci.*, 7, 161, 2023.
11. Bledzki, A.K., Seidlitz, H., Goracy, K., Urbaniak, M., Rösch, J.J., Recycling of carbon fiber reinforced composite polymers—review—part 1: Volume of production, recycling technologies, legislative aspects. *Polymers (Basel).*, 13, 1–13, 2021.
12. Ruane, K., Soutsos, M., Huynh, A., Zhang, Z., Nagle, A., McDonald, K., Gentry, T.R., Leahy, P., Bank, L.C., Construction and Cost Analysis of BladeBridges Made from Decommissioned FRP Wind Turbine Blades. *Sustain.*, 15, 3366, 2023.

13. Shahid, M.K., Mainali, B., Rout, P.R., Lim, J.W., Aslam, M., Al-Rawajfeh, A.E., Choi, Y., A Review of Membrane-Based Desalination Systems Powered by Renewable Energy Sources. *Water (Switzerland).*, 15, 534, 2023.
14. Bahaghighat, M., Xin, Q., Motamedi, S.A., Zanjireh, M.M., Vacavant, A., Estimation of wind turbine angular velocity remotely found on video mining and convolutional neural network. *Appl. Sci.*, 10, 3544, 2020.
15. Haces-Fernandez, F.: Gowind, Wind energy spatiotemporal assessment and characterization of end-of-life activities. *Energies,* 13, 6015, 2020.
16. Luengo, M.M., Kolios, A., Failure mode identification and end of life scenarios of offshore wind turbines: A review. *Energies,* 8, 8339–8354, 2015.
17. Alshahrani, H., Ahmed, A., Kabrein, H., Prakash, V.R.A., Mechanical Properties Study on Sandwich Composites of Glass Fiber Reinforced Plastics (GFRP) Using Liquid Thermoplastic Resin, Elium®: Preliminary Experiments. *Coatings.*, 12, 1423, 2022.
18. Draudvilienė, L., Meškuotienė, A., Raišutis, R., Griškevičius, P., Stasiškienė, Ž., Žukauskas, E., The Usefulness and Limitations of Ultrasonic Lamb Waves in Preventing the Failure of the Wind Turbine Blades. *Appl. Sci.*, 12, 1733, 2022.
19. Ratner, S., Gomonov, K., Revinova, S., Lazanyuk, I., Eco-design of energy production systems: The problem of renewable energy capacity recycling. *Appl. Sci.,* 10, 1–29, 2020.
20. Zou, L., Cheng, H., Research on Wind Turbine Blade Surface Damage Identification Based on Improved Convolution Neural Network. *Appl. Sci.,* 12, 9338, 2022.
21. Sharma, V.B., Singh, K., Gupta, R., Joshi, A., Dubey, R., Gupta, V., Bharadwaj, S., Zafar, M.I., Bajpai, S., Khan, M.A., Srivastava, A., Pathak, D., Biswas, S., Review of structural health monitoring techniques in pipeline and wind turbine industries. *Appl. Syst. Innov.,* 4, 59, 2021.
22. Andersen, N., Eriksson, O., Hillman, K., Wallhagen, M., Wind turbines' end-of-life: Quantification and characterisation of future waste materials on a national level. *Energies,* 9, 999, 2016.
23. Rahimi, B., Nadri, H., Afshar, H. L., Timpka, T., A systematic review of the technology acceptance model in health informatics. *Appl. Clin. Info.*, 9, 604-634, 2018.
24. Abadie, L.M., Goicoechea, N., Old wind farm life extension vs. Full repowering: A review of economic issues and a stochastic application for spain. *Energies,* 14, 3678, 2021.
25. Wei, D., Dye sensitized solar cells. *Int. J. Mol. Sci.,* 11, 1103–1113, 2010.
26. Doerffer, K., Bałdowska-Witos, P., Pysz, M., Doerffer, P., Tomporowski, A., Manufacturing and recycling impact on environmental life cycle assessment of innovative wind power plant part 1/2. *Materials (Basel).*, 14, 1–21, 2021.
27. Olabi, A.G., Wilberforce, T., Elsaid, K., Sayed, E.T., Salameh, T., Abdelkareem, M.A., Baroutaji, A., A review on failure modes of wind turbine components. *Energies,* 14, 5241, 2021.

28. Paulsen, E.B., Enevoldsen, P., A multidisciplinary review of recycling methods for end-of-life wind turbine blades. *Energies*, 14, 1–13, 2021.
29. Pakenham, B., Ermakova, A., Mehmanparast, A., A Review of Life Extension Strategies for Offshore Wind Farms Using Techno-Economic Assessments. *Energies*, 14, 1936, 2021.
30. Mishnaevsky, L., Branner, K., Petersen, H.N., Beauson, J., McGugan, M., Sørensen, B.F., Materials for wind turbine blades: An overview. *Materials (Basel).*, 10, 1–24, 2017.
31. Velenturf, A.P.M., A framework and baseline for the integration of a sustainable circular economy in offshore wind. *Energies*, 14, 5540, 2021.
32. Torres-Madroñero, J.L., Alvarez-Montoya, J., Restrepo-Montoya, D., Tamayo-Avendaño, J.M., Nieto-Londoño, C., Sierra-Pérez, J., Technological and operational aspects that limit small wind turbines performance. *Energies*, 13, 1–39, 2020.
33. Haces-Fernandez, F., Cruz-Mendoza, M., Li, H., Onshore Wind Farm Development: Technologies and Layouts. *Energies*, 15, 2381, 2022.
34. Krauklis, A.E., Karl, C.W., Gagani, A.I., Jørgensen, J.K., Composite material recycling technology—state-of-the-art and sustainable development for the 2020s. *J. Compos. Sci.*, 5, 28, 2021.
35. Rahimizadeh, A., Kalman, J., Henri, R., Fayazbakhsh, K., Lessard, L., Recycled glass fiber composites fromwind turbine waste for 3D printing feedstock: effects of fiber content and interface on mechanical performance. *Materials (Basel).*, 12, 3929, 2019.
36. Im, H., Kim, B., Power Performance Analysis Based on Savonius Wind Turbine Blade Design and Layout Optimization through Rotor Wake Flow Analysis. *Energies*, 15, 9500, 2022.
37. Ngoc, D.M., Luengchavanon, M., Anh, P.T., Humphreys, K., Techato, K., Shades of Green: Life Cycle Assessment of a Novel Small-Scale Vertical Axis Wind Turbine Tree. *Energies*, 15, 7530, 2022.
38. Chatziparaskeva, G., Papamichael, I., Voukkali, I., Loizia, P., Sourkouni, G., Argirusis, C., Zorpas, A.A., End-of-Life of Composite Materials in the Framework of the Circular Economy. *Microplastics.*, 1, 377–392, 2022.
39. Kavaliauskas, Ž., Kėželis, R., Grigaitienė, V., Marcinauskas, L., Milieška, M., Valinčius, V., Uscila, R., Snapkauskienė, V., Gimžauskaitė, D., Baltušnikas, A., Recycling of Wind Turbine Blades into Microfiber Using Plasma Technology. *Materials (Basel).*, 16, 3089, 2023.
40. Elhenawy, Y., Fouad, Y., Marouani, H., Bassyouni, M., Simulation of glass fiber reinforced polypropylene nanocomposites for small wind turbine blades. *Processes*, 9, 1–16, 2021.
41. Castaldo, R., Falco, F. De, Avolio, R., Bossanne, E., Fernandes, F.C., Cocca, M., Pace, E. Di, Errico, M.E., Spinelli, D., Urios, S.A., polymers Critical Factors for the Recycling of Di ff erent End-of-Life Materials : Wood Wastes, Automotive Turbine Blades. *Polymers (Basel).*, 11, 1604, 2019.

42. Pathak, S.M., Kumar, V.P., Bonu, V., Mishnaevsky, L., Lakshmi, R. V., Bera, P., Barshilia, H.C., Development of Cellulose-Reinforced Polyurethane Coatings: A Novel Eco-Friendly Approach for Wind Turbine Blade Protection. *Energies,* 16, 1730, 2023.
43. Piotrowska, K., Piasecka, I., Specification of environmental consequences of the life cycle of selected post-production waste of wind power plants blades. *Materials (Basel).,* 14, 4975, 2021.
44. Mishnaevsky, L., Sustainable end-of-life management of wind turbine blades: Overview of current and coming solutions. *Materials (Basel).,* 14, 1–26, 2021.
45. Jureczko, M., Mrówka, M., Multiobjective Optimization of Composite Wind Turbine Blade. *Materials (Basel).,* 15, 4649, 2022.
46. Gonçalves, R.M., Martinho, A., Oliveira, J.P., Recycling of Reinforced Glass Fibers Waste: Current Status. *Materials (Basel).,* 15, 1–18, 2022.
47. Caterino, N., Pugliano, G., Spizzuoco, M., Robustelli, U., Variable dampers to mitigate structural demand to wind turbines: The role of the monitoring system features for the effectiveness of the control strategy. *Appl. Sci.,* 10, 2498, 2020.
48. Yildiz, N., Hemida, H., Baniotopoulos, C., Maintenance and End-of-Life Analysis in LCA for Barge-Type Floating Wind Turbine. *Wind,* 2, 246–259, 2022.
49. Kasner, R., Kruszelnicka, W., Bałdowska-Witos, P., Flizikowski, J., Tomporowski, A., Sustainable wind power plant modernization. *Energies,* 13, 1461, 2020.
50. Schoden, F., Siebert, A., Keskin, A., Herzig, K., Straus, M., Schwenzfeier-Hellkamp, E., Building a wind power plant from scrap and raising public awareness for renewable energy technology in a circular economy. *Sustain.,* 12, 1–11, 2020.
51. Medici, P., van den Dobbelsteen, A., Peck, D., Safety and health concerns for the users of a playground, built with reused rotor blades from a dismantled wind turbine. *Sustain.,*12, 1–25, 2020.
52. An, J., Zou, Z., Chen, G., Sun, Y., Liu, R., Zheng, L., An IoT-based life cycle assessment platform of wind turbines. *Sensors (Switzerland).,* 21, 1–21, 2021.
53. Parvaresh, A., Abrazeh, S., Mohseni, S.R., Zeitouni, M.J., Gheisarnejad, M., Khooban, M.H., A novel deep learning backstepping controller-based digital twins technology for pitch angle control of variable speed wind turbine. *Designs.,* 4, 1–19, 2020.
54. Xie, R., Chen, M., Liu, W., Jian, H., Shi, Y., Digital twin technologies for turbomachinery in a life cycle perspective: A review. *Sustain.,* 13, 1–21, 2021.
55. Leader, A., Gaustad, G., Critical Material Applications and Intensities in Clean Energy Technologies. *Clean Technol.,* 1, 164–183, 2019.
56. Bashir, S., Thakur, A., Lgaz, H., Chung, I.M., Kumar, A., Corrosion inhibition efficiency of bronopol on aluminium in 0.5 M HCl solution: Insights from experimental and quantum chemical studies. *Surf. and Inter.,* 20, 100542, 2020.

57. Bashir, S., Thakur, A., Lgaz, H., Chung, I.-M., Kumar, A., Corrosion Inhibition Performance of Acarbose on Mild Steel Corrosion in Acidic Medium: An Experimental and Computational Study. *Arab. J. Sci. Eng.*, 45, 4773–4783, 2020.
58. Thakur, A., Kaya, S., Kumar, A., Recent Innovations in Nano Container-Based Self-Healing Coatings in the Construction Industry. *Curr. Nanosci.*, 18, 203–216, 2021.
59. Thakur, A., Kumar, A., Recent trends in nanostructured carbon-based electrochemical sensors for the detection and remediation of persistent toxic substances in real- time analysis. *Mater. Res. Express.*, 10, 034001, 2023.
60. Dhonchak, C., Agnihotri, N., Computational Insights in the Spectrophotometrically 4H-chromen-4-one Complex using DFT Method. *Biointerface Res. Appl. Chem.*, 13, 357, 2023.
61. Bashir, S., Lgaz, H., Chung, I.M., Kumar, A., Effective green corrosion inhibition of aluminium using analgin in acidic medium: an experimental and theoretical study. *Chem. Eng. Commun.*, 0, 1–10, 2020.
62. Thakur, A., Kumar, A., Sustainable Inhibitors for Corrosion Mitigation in Aggressive Corrosive Media: A Comprehensive Study. *J. Bio- Tribo-Corrosion.*, 7, 1–48, 2021.
63. Parveen, G., Bashir, S., Thakur, A., Saha, S.K., Banerjee, P., Kumar, A., Experimental and computational studies of imidazolium based ionic liquid 1-methyl- 3-propylimidazolium iodide on mild steel corrosion in acidic solution Experimental and computational studies of imidazolium based ionic liquid 1-methyl- 3-propylimidazolium. *Mater. Res. Express.*, 7, 016510, 2020.
64. Kumar, A., Overview of the Properties, Applicability, and Recent Advancements of Some Natural Products Used as Potential Inhibitors in Various Corrosive Systems. In: Handbook of Research on Corrosion Sciences and Engineering. pp. 275–277, 2003.
65. Thakur, A., Kumar, A., Zhang, R., Alcoholic Beverage Purification Applications of Activated Carbon. In: Verma, C. and Quraishi, M.A. (eds.) Activated Carbon: Progress and Applications. p. 0. The Royal Society of Chemistry, 2023.
66. Thakur, A., SAVAŞ, K., Kumar, A., Recent Trends in the Characterization and Application Progress of Nano-Modified Coatings in Corrosion Mitigation of Metals and Alloys. *Appl. Sci.*, 13, 730, 2023.
67. Thakur, A., Kaya, S., Abousalem, A.S., Sharma, S., Ganjoo, R., Assad, H., Kumar, A., Computational and experimental studies on the corrosion inhibition performance of an aerial extract of Cnicus Benedictus weed on the acidic corrosion of mild steel. *Process Saf. Environ. Prot.*, 161, 801–818, 2022.
68. Verma, C., Thakur, A., Ganjoo, R., Sharma, S., Assad, H., Coordination bonding and corrosion inhibition potential of nitrogen-rich heterocycles : Azoles and triazines as specific examples. *Coord. Chem. Rev.*, 488, 215177, 2023.

69. Sharma, D., Thakur, A., Om, H., Synthesis, Electrochemical, Morphological, Computational and Corrosion Inhibition Studies of 3-(5-Naphthalen-2-yl-[1,3,4]oxadiazol-2-yl)-pyridine against Mild Steel in 1 M HCl. *Asian J. Chem.*, 35, 1079–1088, 2014.
70. Kaya, S., Thakur, A., Kumar, A., The role of in Silico / DFT investigations in analyzing dye molecules for enhanced solar cell efficiency and reduced toxicity. *J. Mol. Graph. Model.*, 124, 108536, 2023.
71. Thakur, A., Kumar, A., Recent advances on rapid detection and remediation of environmental pollutants utilizing nanomaterials-based (bio)sensors. *Sci. Total Environ.*, 834, 155219, 2022.
72. Thakur, A., Kaya, S., Abousalem, A.S., Kumar, A., Experimental , DFT and MC simulation analysis of Vicia Sativa weed aerial extract as sustainable and eco-benign corrosion inhibitor for mild steel in acidic environment. *Sustain. Chem. Pharm.*, 29, 100785, 2022.
73. Thakur, A., Kumar, A., Sharma, S., Ganjoo, R., Assad, H., Materials Today : Proceedings Computational and experimental studies on the efficiency of Sonchus arvensis as green corrosion inhibitor for mild steel in 0 . 5 M HCl solution. *Mater. Today Proc.*, 66, 609–621, 2022.
74. Thakur, A., Kumar, A., Kaya, S., Marzouki, R., Zhang, F., Guo, L., Recent Advancements in Surface Modification , Characterization and Functionalization for Enhancing the Biocompatibility and Corrosion Resistance of Biomedical Implants. *Coatings.*, 12, 1459, 2022.
75. Thakur, A., Sharma, S., Ganjoo, R., Assad, H., Kumar, A., Anti-Corrosive Potential of the Sustainable Corrosion Inhibitors Based on Biomass Waste: A Review on Preceding and Perspective Research. *J. Phys. Conf. Ser.*, 2267, 012079, 2022.
76. Thakur, A., Kumar, A., Kaya, S., Vo, D.V.N., Sharma, A., Suppressing inhibitory compounds by nanomaterials for highly efficient biofuel production: A review. *Fuel.*, 312, 122934, 2022.
77. Kumar, A., Thakur, A., Encapsulated nanoparticles in organic polymers for corrosion inhibition. Elsevier Inc., 2020.
78. Bashir, S., Thakur, A., Lgaz, H., Chung, I.-M., Kumar, A., Computational and experimental studies on Phenylephrine as anti-corrosion substance of mild steel in acidic medium. *J. Mol. Liq.*, 293, 111539, 2019.
79. Abdallah, R., Juaidi, A., Savaş, M.A., Çamur, H., Albatayneh, A., Abdala, S., Manzano-Agugliaro, F., A critical review on recycling composite waste using pyrolysis for sustainable development. *Energies*, 14, 2021.
80. Nie, Y., Liu, Q., Xiang, Z., Zhong, S., Huang, X., Performance and Modification Mechanism of Recycled Glass Fiber of Wind Turbine Blades and SBS Composite-Modified Asphalt. *Appl. Sci.*, 13, 6335, 2023.
81. Yildiz, N., Hemida, H., Baniotopoulos, C., Life cycle assessment of a barge-type floating wind turbine and comparison with other types of wind turbines. *Energies*, 14, 5656, 2021.

82. Rajpar, A.H., Ali, I., Eladwi, A.E., Bashir, M.B.A., Recent development in the design of wind deflectors for vertical axis wind turbine: A review. *Energies*, 14, 5140, 2021.
83. Taras, P., Nilifard, R., Zhu, Z.Q., Azar, Z., Cooling Techniques in Direct-Drive Generators for Wind Power Application. *Energies*, 15, 5986, 2022.
84. Koumoulos, E.P., Trompeta, A.F., Santos, R.M., Martins, M., Dos Santos, C.M., Iglesias, V., Böhm, R., Gong, G., Chiminelli, A., Verpoest, I., Kiekens, P., Charitidis, C.A., Research and development in carbon fibers and advanced high-performance composites supply chain in Europe: A roadmap for challenges and the industrial uptake. *J. Compos. Sci.*, 3, 86, 2019.
85. Hemeida, M.G., Hemeida, A.M., Senjyu, T., Osheba, D., Renewable Energy Resources Technologies and Life Cycle Assessment: Review. *Energies*, 15, 1–36, 9417, 2022.
86. Ren, Z., Wu, L., Pang, Y., Zhang, W., Yang, R., Strategies for effectively harvesting wind energy based on triboelectric nanogenerators. *Nano Energy.*, 100, 245–268, 2022.
87. Mamun, K.A., Islam, F.R., Haque, R., Chand, A.A., Prasad, K.A., Goundar, K.K., Prakash, K., Maharaj, S., Systematic Modeling and Analysis of On-Board Vehicle Integrated Novel Hybrid Renewable Energy System with Storage for Electric Vehicles. *Sustain.*, 14, 1–33, 2022.
88. Flizikowski, J., Kruszelnicka, W., Macko, M., The development of efficient contaminated polymer materials shredding in recycling processes. *Polymers (Basel).*, 13, 1–44, 2021.
89. Feng, Y., Zhang, L., The GHG Intensities of Wind Power Plants in China from a Life-Cycle Perspective: The Impacts of Geographical Location, Turbine Technology and Management Level. *Sustain.*, 15, 4449, 2023.
90. Korniejenko, K., Kozub, B., Bąk, A., Balamurugan, P., Uthayakumar, M., Furtos, G., Tackling the circular economy challenges—composites recycling: Used tyres, wind turbine blades, and solar panels. *J. Compos. Sci.*, 5, 1–18, 2021.
91. Oelker, S., Sander, A., Kreutz, M., Ait-Alla, A., Freitag, M., Evaluation of the impact of weather related limitations on the installation of offshore wind turbine towers. *Energies*, 14, 1–12, 2021.

8

Wind Turbine Assessment: A Step-by-Step Approach

Figen Balo[1,2]* and Lutfu S. Sua[2]

[1]*Department of METE, Engineering Faculty, Firat University, Elazig, Turkey*
[2]*Department of Management and Marketing, Southern University and A&M College, Baton Rouge, LA, USA*

Abstract

The weighted sum approach is suggested in this work as a multi-attribute decision-support strategy for the wind turbine selection problem. The inherent constraints of earlier investigations served as the inspiration for the suggested approach. These restrictions resulted from the use of computationally challenging methodologies, a lack of diversity of wind turbines, and simple decision models with a single criterion. Analytic hierarchy strategy is a methodology that is utilized to overcome various decision-making difficulties. Because it takes into account both intangible and tangible attributes, the analytic hierarchy strategy is popular for providing answers to multi-criteria problems. This strategy is used in the study to predict the factor weights and determine the elements' relative weights involved in the wind turbine evaluation. Application of the suggested strategy to several types of wind turbines from various manufacturers is used for an analysis of the effectiveness of the approach. This research considers the participation of major elements in selecting a wind turbine for a wind facility in Turkey in order to build sustainable development management solutions. In this regard, an analytic hierarchy strategy was used to balance and choose the best alternative for the criteria and sub-criteria after calculating the significant aspects in the evaluation of a wind turbine technology in accordance with the principles of sustainable development management. Based on the findings, the most optimum alternative is determined as the top choice.

Keywords: MCDM, renewable energy, wind turbine selection, wind energy, electric sector's sustainable development

**Corresponding author*: figenbalo@gmail.com

Inamuddin, Tariq Altalhi and Mohammad Luqman (eds.) Wind Energy Storage and Conversion: From Basics to Utilities, (185–206) © 2024 Scrivener Publishing LLC

8.1 Introduction

In order to attain sustainability in any field, sustainable development refers to a method that calls for an integrated and prompt replacement with the use of renewable resources [1]. Many researchers have been interested in sustainable development in recent years [2], and it has been suggested as a new strategy so that communities might profit from increased living standards, social justice, and resource conservation [3]. Today the theory of sustainable development, which dates back to the last decades of the twentieth century, has already gained strength. According to studies, communities' energy supply will be dominated by new and renewable energy sources in the near future. For developing nations with arid climates, renewable energy is particularly alluring due to its high sustainability, non-pollution, ease of use, and similar factors [4]. During the last several decades, the relevance of sustainable power sources has increased owing to the world's growing energy demand and the depletion of fossil-based energy resources. Sustainable power sources like solar power and wind power should be employed more frequently to reduce the ultimate overall energy consumption from fossil fuels. Wind power is particularly attractive because of the fact that wind can perpetually produce energy 24 h a day [5]. Because wind energy does not emit any greenhouse gases or pollution, it is referred to as clean electricity [6].

Solar power is used to create wind power that is used through wind turbines. The earth's surface's uneven cooling and warming result in wind generation. The sun's radiation penetrates farther into the atmosphere in the polar areas than it does in the equatorial zones due to the roundness of the planet and the mutual location of the earth and the sun, resulting in temperature differences. As a result of these areas of low and high pressure trying to balance each other out, prevailing winds, which are large-scale air currents, are created. Local winds are winds that cover confined and more compact areas. Numerous variables, including variations in land–sea temperatures, geography, landforms, and terrain coverage, have an impact on local winds [7].

The benefits of wind energy over conventional power production technologies (nuclear, gas, coal facilities, etc.) lay in the commissioning and rapid deployment of wind facilities. This is attributable to wind turbines, which need little effort and expense in upkeep and operation. Additionally, wind turbines have an operational lifespan of 20 to 25 years, which is very cost-effective. In addition, there are no geographical restrictions on harnessing wind energy, which is not the case with fossil fuel-sourced energy generation mechanisms [8, 9]. When wind turbines convert the kinetic power of the wind into electric power, electricity is created. The wind propels the turbine blades into rotation when it reaches a particular speed, often 3 to 4 m/s. The generator is turned by the blades in the machinery room or nacelle at the top of the tower. This generates electricity that is subsequently put into the system and delivered to the final users after being transformed into grid voltage by a transformer.

Wind energy has become one of the most rapidly expanding sustainable energy resources as countries seek more of renewable and sustainable energy resources owing to the concerns about energy security. When the stakes are high and a choice must be made quickly, it is crucial to use a methodical framework that can quantitatively deal with operational relationships in a complicated system. By logically fusing judgment and personal values, the analytical hierarchy technique is both a strong organizing tool and a model for setting priorities. For this reason, it is a method that gives meaningful results in the assessment of sustainable energy systems for different purposes. Some multi-criteria studies carried out for this purpose are given below.

In order to evaluate wind power plants, Ciraolo and Cavallaro applied the fuzzy-new F-NA-IADE strategy to decision contexts and fuzzy assessment. Four different plans were used to come up with an alternative. Aesthetics, noise, environmental impact, and social acceptability were utilized as attributes along with investment costs, operating and maintenance expenses, energy generation capacities, energy savings, technic advancement, CO_2 adsorption, and materialization times. To evaluate the changes in the order, a sensitivity analysis was also carried out to evaluate the results' robustness [10]. Romulo, Eudemario, and Alex implemented an analytic hierarchy strategy for a wind farm project with multiple criteria. Based on the values of the wind speed, the analytic hierarchy strategy technique ranked and categorized all 27 territories for the installation of the wind farms [11]. Birol and Ozgur used an analytic hierarchy strategy for power source design to create a district's environment using multiple levels of criteria. Residents and the local government supported sun power investments, but the sector chose to finance cleaner technologies in decentralized lignite energy facilities and geothermal energy facilities [12]. The decision-making processes employed by Kolios *et al.* for the wind turbine support structure choice. Their research compared the PROMETHEE and ELECTRE, AHP, WPM, WSM, and TOPSIS approaches. Utilizing stochastic algorithms created with a Monte Carlo methodology, the best outcomes were discovered [13]. The subject of wind turbine selection has drawn a lot of attention over the last 20 years. In literature, a range of variants in wind turbine selection problems has been recorded [14–18]. A system reliability-based approach was created by Munteanu and Nemes to evaluate nine different turbine designs [19]. To determine the most effective wind turbine, Perkin *et al.* used a genetic algorithm and a variety of selection parameters, including the generator size, rotor radius, pitch angle, and hub height [20]. The research of Sua and Balo for the choice of a 1.5-MW wind turbine was another study that utilized the AHP. The selected criteria included the system's maximal capacity, rotor diameter, energy output, hub height, overall expense, noise level, state support, electromagnetic effect, integration potential, and degree of satisfaction [21]. In a probabilistic model that they created, Dobakhshari and Firuzabad employed turbine reliability as the selection criterion. Five different types of turbines were used to test the suggested strategy [22]. When conducting interviews with subject matter experts, Halonen and Sarja used a qualitative approach. Their study revealed a number of turbine selection variables, including product availability and dependability, price, vendor production frequency, and maintenance trends [23]. Chowdhury *et al.* suggested a particle swarm optimization-based technique for choosing a turbine [24]. They took into consideration a sole wind turbine type and used power output capacity as their deciding factor. More than 120 different wind turbine types were taken into consideration, utilizing the expense of energy as the wind turbine selection criterion in a method suggested by Chowdhury [25]. To choose the optimal wind turbine, Montoya *et al.* devised a genetic methodology based on Pareto ranking. The selection criteria in their decision modeling included daily energy output variance and power output [26]. Bencherif *et al.* devised an analytical strategy depending on the Weibull distribution and took into account 24 diverse wind turbine models utilizing the capacity factor as the attribute for decision-making [27]. Martin made a fictitious wind turbine as the starting point and created a straightforward support tool while considering a variety of wind situations [28]. Helgason researched a number of probable Icelandic locations. The cost of electricity was utilized as the criterion for choosing the turbines, and 47 various turbine modeling options were taken into consideration [29]. Blade design was used as the deciding factor in Ramayya and Bekele's consideration of a site-specific turbine selection. To improve their model, they suggested using a genetic algorithm [30]. Onyewudiala and Eke suggested a genetic methodology for site-specific wind turbine selection [31]. The optimization model used blade thickness, cord, and twist with the goal of maximizing power generation. Six different turbine models were employed in Aljowder's process for selecting a turbine, and the capacity factor was used as the deciding element [32]. Jureczko *et al.* used a genetic algorithm to design turbines while taking into account certain design goals. These goals included generated output, necessary blade strength, cost of the blade material, vibrations of the blade, and stability of the blade construction [33]. With turbine efficiency index, capacity factor, and mean energy output as the deciding factors, El Shimy suggested an area-specific wind turbine choice technique [34]. Shirgholami *et al.* developed an analytical hierarchy strategy-based approach and found over 30 decision criteria. However, relying on the site-specific circumstances, only a fraction of these attributes might be utilized in the selection procedure [35, 36]. Turbine choice was taken into account by Dong *et al.* while taking wind turbine expense and hybrid matching displays into consideration as the optimization attribute. Differential evaluation, particle swarm optimization, and the proposed modeling were all used [37]. A technique based on weighted sums was suggested by Bagočius *et al.* for choosing the turbines for offshore wind farms. They took into account five aspects when making their decision: maximal power produced in the region, annual energy production, CO_2 emissions, investments, and nominal wind turbine power [38]. Du *et al.* developed a wind turbine selection strategy depending on SCADA information analysis [39]. Rehman and Khan employed a multi-attribute decision-support strategy depending on fuzzy logic and three attributes to suggest a turbine selection approach [40, 41]. Environmental considerations, financial considerations, machine features, and technical difficulties were included as the main decision factors in the proposal of Lee *et al.* of a multi-attribute decision strategy. They looked at four turbines, each with roughly the same rated power [42]. They then added a fuzzy logic-sourced wind turbine selection technique with six factors to their proposal [43]. A summary of the utilized multi-criteria methodologies in wind turbine selection is given in Table 8.1.

Managers must act wisely when deciding which wind turbines to place in each project because just a few wind turbine firms have gained an existence in international markets. It is challenging to solve the wind turbine selection problem with significant uncertainty using conventional methods since it combines both quantitative and qualitative variables. In order to create a practical and useful tool for decision-makers, the assessment criteria system for wind turbines is built by carefully taking into account a variety of indexes. This study suggests a decision framework based on the analytic hierarchy strategy to address the complexity of the choice environment and the correlation among criteria. This methodology is a series of properties that were chosen to obtain the notional significance. To produce a more practical set of attributes, this research examines the literature in choosing wind turbines, seeks input from professionals in the field, and the most efficient wind turbine determined according to wind turbine criteria and alternatives by an analytical hierarchy technique.

Table 8.1 Multi-criteria methodologies in wind turbine selection.

Ref.	Methods										
	GP	WPM	PROMETHEE	ELECTRE	TOPSIS	F-NAIADE	WASPAS	F-ANP	WSM	F-LOGIC	AHP
[10]						1					
[42]								1			
[44]							1				1
[8]										1	
[36]											1
[43]										1	
[13]		1	1	1	1				1		1
[45]									1		
[21]											1
[46]	1	1	1	1	1	1	1	1	2	2	4

8.2 Analytic Hierarchy Strategy

A collection of methods known as multi-attribute decision-making are used to compare, rank, and choose between options based on both quantitative and non-quantitative factors. Multi-attribute decision-making was developed to address a variety of issues. One of these issues is the selection problem, which is the subject of our review. The most efficient alternative from a list of options is selected using multi-attribute decision-making. The analytic hierarchy strategy is a methodology for assessing and organizing complicated decisions that are based on psychology and math. Saaty introduced this method in the 1970s, and it has subsequently been further improved. It is separated into three parts: the eventual problem or goal that it is aiming to discuss, overall feasible solutions (also called options), and the attribute it will utilize to assess the options. A multi-attribute decision-support methodology's mission is to examine several options in the context of various criteria and competing objectives [47]. The international scientific community has acknowledged the analytic hierarchy strategy as an adaptable and reliable multi-attribute decision-support tool for analyzing complex choice issues [48].

The analytic hierarchy strategy's disadvantages are as follows [49–52]:

- With all the subsystems required for problem reversal, an analytic hierarchy strategy can take a while, especially if there are a lot of criteria and options—for example, to establish the weights for the 10- criteria hierarchy, the decision maker must perform 45 pairwise comparison calculations.
- Analytic hierarchy strategy can be thought of as a series of processes, and the aggregation of these steps might lead to the loss of crucial information. It can be difficult to compensate for good and negative criterion scores.
- Due to a phenomenon known as rank reversal, the analytic hierarchy strategy is susceptible to ranking abnormalities like many MCDM approaches. When a comparable option is included in the list of alternatives being considered, this happens. This makes it difficult to interpret the weights of the criteria.

On the positive side, the method:

- allows people to clarify and record their issues, and through repetition, their judgment is improved.

- combines a representative verdict from a range of opinions.
- helps people to choose the optimal option depending on their goals while taking into account the relative priority of many elements.
- leads to a determination of how desirable each choice is.
- keeps track of the reasoning of judgments.
- gives a scale for evaluating priorities and intangibles.
- complies with the natural tendency of humans to divide systems into levels and elements.
- integrates deductive reasoning and inductive reasoning, two key human analytical methods, into a coherent framework.
- is adaptable and simple to understand for a variety of issues.

Four previously mentioned criteria are used to construct the measurement of flexibility. Each criterion in the measurement of case flexibility has its own importance. The option should be broken down into subsequent phases in order to organize the criteria based on their impact on flexibility [53], namely:

- Define the issue, identify the specifications that are known to exist, and create a hierarchy that goes from the aim of the primary four attributes to the lesser level of options.
- Create a series of binary crosscheck matrices for each criterion in relation to the primary target.
- Utilize the preferences to give each alternative and each criterion equal weights.
- Order the options in accordance with the objective of assessing the alternatives' weights.

Imagine that n changes, $A_1,.....,A_n$, whose weights $w_1,.....,w_n$, are acknowledged, respectively. The weight ratio in accordance with the fundamental scale is contained in the binary matrix. According to the reciprocal of the weights, a comparison is made between each pair of alternatives, as shown in the matrix equation below:

$$\begin{pmatrix} \omega_1/\omega_1 & \omega_1/\omega_2 & \cdots & \omega_1/\omega_n \\ \omega_2/\omega_1 & \omega_2/\omega_2 & \cdots & \omega_2/\omega_n \\ \vdots & \vdots & \vdots & \vdots \\ \omega_n/\omega_1 & \omega_n/\omega_2 & \cdots & \omega_n/\omega_n \end{pmatrix} \begin{pmatrix} \omega_1 \\ \omega_2 \\ \vdots \\ \omega_n \end{pmatrix} = n \begin{pmatrix} \omega_1 \\ \omega_2 \\ \vdots \\ \omega_n \end{pmatrix} \tag{8.1}$$

It can be summed up as follows:

$$A\omega = n\omega \tag{8.2}$$

According to Figure 8.1, the binary crosscheck matrices in the analytic hierarchy strategy are based on the relevance intensity's basic measurement [54].

The analytic hierarchy strategy's random consistency index and fundamental scale are shown in Figure 8.2. The answer to Equation 8.2 is known as the fundamental right eigenvector of A. The elements must be standardized by separating over their total. Priorities must be calculated, and then the alternative must be gauged to ensure consistency. The following consistency index in Equation 8.3 can be used to gauge consistency for solution evaluation.

$$CI = \frac{\lambda_{\max} - n}{n-1} \tag{8.3}$$

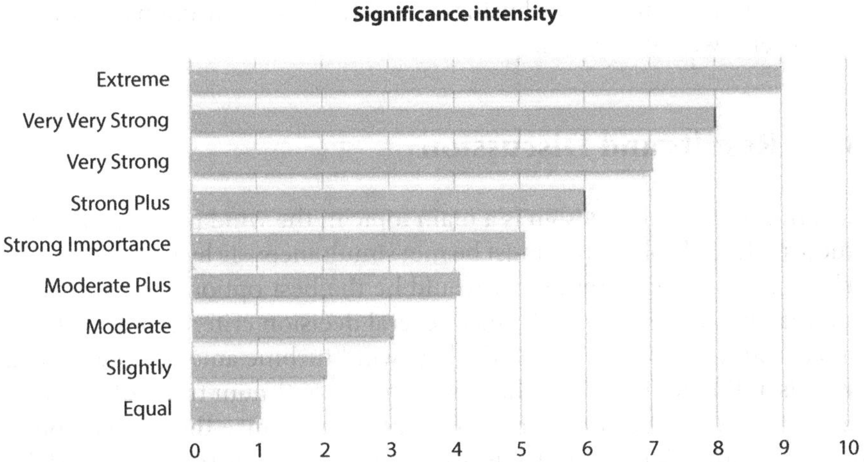

Figure 8.1 Analytic hierarchy strategy random consistency index and fundamental scale.

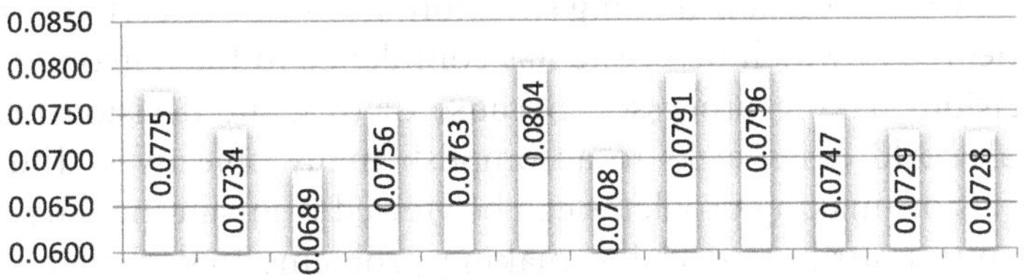

Figure 8.2 Total scores of the alternatives.

where λ_{max} is the eigenvector's biggest primary eigenvalue. It is necessary to contrast the consistency index (CI) with the pre-calculated random consistency index (RI) [55]. The consistency ratio is then determined as in Equation 8.4:

$$CR = \frac{CI}{RI} \tag{8.4}$$

The inconsistency is acceptable if the consistency rate value is less than or equal to 0.1, while the ratio in Equation 8.4 will be applied to each alternative's comparison of the criteria to the goal and to the criteria.

The required data was gathered in accordance with the model's design and the various levels of the analytic hierarchy strategy tree in this study in order to establish the analytic hierarchy strategy, and the weighting process was then carried out based on this. In order to meet the criteria and sub-conditions, several interviews were done. Additionally, a questionnaire created by a researcher in this area was used and given to those working in the wind turbine business. Finally, the agreed-upon criteria and sub-criteria were computed.

8.3 Results and Discussion

Effective wind facility design is a main topic in the wind power generation studies. Multiple objectives must be met simultaneously by this design. The selection of a wind turbine that would be the best option for a particular wind facility area is one goal. Since several decision criteria must be taken into account at once, choosing the best wind turbine among a few diverse types is a difficult issue. For this decision, the optimum trade-off between the criteria must be considered in order to maximize the satisfaction of each criterion. Finding and selecting the best option from the available possibilities is the process of making a decision. Due to the existence of diverse and occasionally contradictory criteria, the decision-maker in many real-world decision-making situations has some uncertainty regarding the degree to which he or she has confidence in the outcomes of various decision-making processes. Because of this, the decision-maker in most real-world cases seeks to accomplish many objectives. Thus, real-world decision-making problems can be viewed from another angle in the form of multi-attribute decision-making problems [56]. Several aspects with both good and negative consequences should be considered when selecting a wind turbine. Therefore, the adoption of multidisciplinary decision models is required for appropriate decision-making in this domain. Three basic processes make up multi-attribute decision-making: identifying and assessing the indications, weighing each indicator, and choosing the best alternative. The analytic hierarchy approach was implemented in this article to weigh each indicator and choose the best choice after identifying and assessing the indicators. All of the criteria for choosing the most effective wind turbine technology in this research were broken down into four categories for this reason. These four significant attributes for benchmarking technologies are generator (voltage), power (cut-in wind speed, cut-out wind speed, and rated wind speed), rotor (swept area, diameter, and power density 1 and power density 2), and tower (max hub height). After performing in-depth research, the method of expert interviews was utilized in the current study to enumerate the critical elements in assessing wind turbine technology in accordance with the principles of sustainable development management. To meet the criterion and sub-criteria, multiple interviews were done in this respect. The stakeholders in the wind turbine sector were also given a questionnaire created by the researcher in this area. Finally, the agreed-upon criteria (four attributes) and sub-criteria (nine sub-attributes) were tallied. Based on the country's current position, 12 foreign firms that provide wind turbines for wind farms were recognized as potential international choice options.

Based on nine selection criteria, the attractiveness of 12 potential wind turbines is evaluated using the analytic hierarchy strategy. After selecting the decision criteria and options, creating the questionnaire, consulting the research's statistical community, and taking into account the distribution and interviewing methods, the responses of the chosen respondents were gathered and paired. These responses represented their assessments of the significance and priority of the indicators. There is no need to conduct a validity test because the study questionnaires are, in reality, specialized questionnaires designed to prioritize and count the judgments of two-by-two pairs of experts [57]. The judicial process has been carried out from the bottom up in the current study because evaluating the criteria depends on the options being taken into account. Given that the data analysis method depends on the analytic hierarchy strategy, the incompatibility values of the pairwise comparisons of the respondents were first monitored; after this, the pairwise comparison matrix was calculated with the assurance of an acceptable level of incompatibility (values less than 0.1). The most efficient wind turbine, as determined by the results, was one of the four factors considered in choosing the right wind turbine for wind farms. Table 8.2 presents the selected characteristics of 12 wind turbines

Table 8.2 3MW wind turbine attributes.

	WT1	WT2	WT3	WT4	WT5	WT6	WT7	WT8	WT9	WT 10	WT 11	WT 12
Power												
Rated wind speed	10	10.5	11.5	11	9.5	10	10.5	16	12	12	15	13
Cut-out wind speed	20	25	25	25	22	25	20	34	20	25	25	25
Cut-in wind speed	3	3	3	5	3	3	3	3	3	3	3	3
Rotor												
Diameter	140	122	103	118.2	136	145	119	82	141	120	101	100
Swept area	15,393.8	11,689	8,328	10,973	14.526.7	16,512.9	11,122	5,281	15,164	11,390	8,012	7,850
Power density 1	194.9	256.7	360.2	273.4	206.5	181.7	269.7	568.1	197.8	263.4	374.4	382.2
Power density 2	5.1	3.9	2.8	3.7	4.8	5.5	3.7	1.8	5.1	3.8	2.7	2.6
Generator												
Voltage	690	680	690	1,000	690	690	680	690	690	690	690	860
Tower												
Maximum hub height	110	139	85	140	120	111	112	138	140	160	143	100

compared in this study. The characteristics provided in Table 8.2 are used within the multi-attribute decision-making methodology.

The values in Table 8.2 need to be normalized for further analysis. Table 8.3 presents the normalized values of the 12 wind turbine alternatives.

The normalized values in Table 8.3 need to be multiplied with the priority of each criterion to calculate the weighted score of each turbine alternative. Table 8.4 presents the decision matrix developed by a panel of expert opinions to provide a pairwise comparison of the decision criteria used in the study.

The correlation matrix in Table 8.4 indicates that power density is the most significant factor contributing to the overall attractiveness of solar panels. Table 8.5 presents the weighted scores of the wind turbine alternatives after multiplying the normalized values in Table 8.3 with the weights obtained from Table 8.4.

Figure 8.2 provides a comparison of the alternative wind turbines depending on the nine attributes. The outputs display that wind turbine 8 has the highest score among the 12 brands compared in this study.

8.4 Conclusions

As the environmental effects of fossil fuels become increasingly obvious in recent years, there will be a greater emphasis placed on the need for sustainable energy worldwide. The demand for power generation and delivery is further increased by the rapid population growth in emerging nations. Without a doubt, fossil-based energy sources produce power of high quality, but they are not cost-effective over the long term and then eventually run out. Fossil fuels are also to blame for a number of ecological problems and pose a severe threat to the balance of the environment [58]. Renewable energy is significant because of a lower need for fossil-based power sources to supply cleaner energy and environmental balance topics, etc. Due to the global energy crisis, the majority of nations today have adopted policies and plans that use renewable resources to produce energy. According to a recent International Energy Agency study, it is anticipated that global renewable energy capacity will expand by 50% in the upcoming year and by 2400 GW (or 75% in the following 5 years) [59]. The production of industrial energy is evident to everyone in both the present and future civilizations due to the sustainable and sustainable development of clean energy (like wind energy). The analytic hierarchy strategy of feasible wind turbine intangible qualitative attributes in addition to the quantitative attributes involves a decision that was simpler. This is a well-defined

Table 8.3 Normalized attributes.

	WT1	WT2	WT3	WT4	WT5	WT6	WT7	WT8	WT9	WT10	WT11	WT12
Rated wind speed	0.071	0.074	0.082	0.078	0.067	0.071	0.074	0.113	0.085	0.085	0.106	0.092
Cut-out wind speed	0.069	0.086	0.086	0.086	0.076	0.086	0.069	0.117	0.069	0.086	0.086	0.086
Cut-in wind speed	0.079	0.079	0.079	0.132	0.079	0.079	0.079	0.079	0.079	0.079	0.079	0.079
Diameter	0.098	0.085	0.072	0.083	0.095	0.102	0.083	0.057	0.099	0.084	0.071	0.070
Swept area	0.113	0.086	0.061	0.081	0.107	0.121	0.082	0.039	0.111	0.084	0.059	0.058
Power density 1	0.055	0.073	0.102	0.077	0.059	0.051	0.076	0.161	0.056	0.075	0.106	0.108
Power density 2	0.112	0.086	0.062	0.081	0.105	0.121	0.081	0.040	0.112	0.084	0.059	0.057
Voltage	0.079	0.078	0.079	0.114	0.079	0.079	0.078	0.079	0.079	0.079	0.079	0.098
Maximum hub height	0.073	0.093	0.057	0.093	0.080	0.074	0.075	0.092	0.093	0.107	0.095	0.067

Table 8.4 Correlation matrix.

	Rated wind speed	Cut-out wind speed (m/s)	Cut-in wind speed (m/s)	Diameter (m)	Swept area (m^2)	Power density 1 (W/m^2)	Power density 2	Voltage	Maximum hub height
Rated wind speed	1	0.5	0.33	0.5	2	0.20	0.2	0.33	0.5
Cut-out wind speed		1	0.5	0.33	0.16	0.12	0.12	0.16	0.5
Cut-in wind speed			1	0.20	0.33	0.25	0.25	0.2	0.33
Diameter				1	0.25	0.16	0.16	0.5	2
Swept area					1	0.33	0.33	0.5	2
Power density 1						1	1	2	3
Power density 2							1	2	3
Voltage								1	2
Maximum hub height									1
Weights	4.9%	3.2%	4.9%	8.0%	10.6%	23.6%	23.6%	14.20%	7.0%

Table 8.5 Weighted scores.

	WT1	WT2	WT3	WT4	WT5	WT6	WT7	WT8	WT9	WT10	WT11	WT12
Rated wind speed	0.0035	0.0036	0.0040	0.0038	0.0033	0.0035	0.0036	0.0056	0.0042	0.0042	0.0052	0.0045
Cut-out wind speed	0.0022	0.0027	0.0027	0.0027	0.0024	0.0027	0.0022	0.0037	0.0022	0.0027	0.0027	0.0027
Cut-in wind speed	0.0039	0.0039	0.0039	0.0064	0.0039	0.0039	0.0039	0.0039	0.0039	0.0039	0.0039	0.0039
Diameter	0.0078	0.0068	0.0058	0.0066	0.0076	0.0081	0.0067	0.0046	0.0079	0.0067	0.0057	0.0056
Swept area	0.0120	0.0091	0.0065	0.0085	0.0113	0.0128	0.0087	0.0041	0.0118	0.0089	0.0062	0.0061
Power density 1	0.0130	0.0172	0.0241	0.0183	0.0138	0.0122	0.0180	0.0380	0.0132	0.0176	0.0250	0.0256
Power density 2	0.0265	0.0202	0.0145	0.0192	0.0249	0.0285	0.0192	0.0093	0.0265	0.0197	0.0140	0.0135
Voltage	0.0112	0.0110	0.0112	0.0162	0.0112	0.0112	0.0110	0.0112	0.0112	0.0112	0.0112	0.0140
Maximum hub height	0.0051	0.0065	0.0040	0.0065	0.0056	0.0052	0.0052	0.0064	0.0065	0.0075	0.0067	0.0047
Total	0.0775	0.0734	0.0689	0.0756	0.0763	0.0804	0.0708	0.0791	0.0796	0.0747	0.0729	0.0728

arithmetical formation of a coherent and approximate or accurate weight matrix to compose, depending on the similar true eigenvector ability [60].

In this article, an effort was made to evaluate and choose wind turbines for the conditions in Turkey using multi-attribute judgments based on numerous positive and negative characteristics. In order to do this, elements and indicators of the superiority of a wind turbine technology were tallied and divided into four broad categories based on the findings of the literature and multiple expert interviews. The analytic hierarchy strategy was used in this paper to weigh and choose the best option among the products of 12 international companies selling wind turbines for wind farms after identifying and evaluating the criteria (four criteria) and important sub-criteria (nine sub-criteria) in evaluating and selecting a wind turbine technology. Based on the findings and taking into account all facets of the occasionally contradictory criteria, WT6 was ultimately determined to be the best choice.

The analytic hierarchy strategy, on which the proposed turbine selection methodology is based, has several innovative features [61]. The suggested method creates a turbine selection model while taking into account four straightforward but crucial decision criteria and taking conflict and incommensurability into consideration. Any commercially available turbine can easily and quickly meet these requirements, simplifying the suggested method. The results are more thorough because 12 turbines from different manufacturers have also been taken into account. It is also critical to note that the proposed framework is flexible and reliable; attributes and turbine types can be simply added or removed to fulfill the requirements of the designer without compromising computational effectiveness.

References

1. Potter, R., Conway, D., Evans, R., Lloyd-Evans, S., Sustainable development and environmental sustainability, 2014, http://dx.doi.org/10.4135/9781473914834.n13.
2. Naess, P., Planning for a sustainable and less car-dependent urban development, 2020, http://dx.doi.org/10.4324/9780203099186-20.
3. Cater, E., Ecotourism in the third world: Problems for sustainable tourism development. *Tour. Manage.*, 14, 85, 1993.
4. Rafael, C.H. and Lucas, M.G., Ecosystem vulnerability due to urban growth policies. The role of regulatory ecosystem contributions in the land use planning, 2022, http://dx.doi.org/10.31428/10317/10599.
5. Balo, F. and Sagbansua, L., A multi-criterion analysis for choosing photovoltaic panel. *iJARS Int. J. Eng.*, 2, 2016.

6. Aghajani, D., Abbaspour, M., Radfar, R., Mohammadi, A., Using Web-GIS technology as a smart tool for resiliency management to monitor wind farms performances (Ganjeh site, Iran). *Int. J. Environ. Sci. Technol.*, 16, 5011, 2019.
7. Fingerroos, O., Painavaa tietoa maahanmuuttajanaisista. *Elore*, 16, 2009.
8. Khan, S.A. and Rehman, S., On the use of Unified And-Or fuzzy aggregation operator for multi-criteria decision making in wind farm design process using wind turbines in 500 kW - 750 kW range, 2012, http://dx.doi.org/10.1109/fuzz-ieee.2012.6251306.
9. Rehman, S. and Al-Abbadi, N.M., Wind power characteristics on the North West coast of Saudi Arabia. *Energy Environ.*, 20–21, 1257, 2009.
10. Cavallaro, F. and Ciraolo, L., A multicriteria approach to evaluate wind energy plants on an Italian island. *Energy Policy*, 33, 235, 2005.
11. Bulhões, R.L., De Santana, E.S., Santos, A.Á.B., Use of analytic hierarchy process for wind farm installation region prioritization-case study. *Energies*, 13, 2284, 2020.
12. Erol, Ö. and Kilkiş, B., An energy source policy assessment using analytical hierarchy process. *Energy Convers. Manage.*, 63, 245, 2012.
13. Kolios, A., Mytilinou, V., Lozano-Minguez, E., Salonitis, K., A comparative study of multiple-criteria decision-making methods under stochastic inputs. *Energies*, 9, 566, 2016.
14. Perkin, S., Garrett, D., Jensson, P., Optimal wind turbine selection methodology: A case-study for Búrfell, Iceland. *Renew. Energy*, 75, 165, 2015.
15. Nemes, C. and Munteanu, F., Optimal selection of wind turbine for a specific area, 2010, http://dx.doi.org/10.1109/optim.2010.5510577.
16. Chowdhury, S., Zhang, J., Messac, A., Castillo, L., Optimizing the arrangement and the selection of turbines for wind farms subject to varying wind conditions. *Renew. Energy*, 52, 273, 2013.
17. Dobakhshari, A.S. and Fotuhi-Firuzabad, M., A reliability model of large wind farms for power system adequacy studies. *IEEE Trans. Energy Convers.*, 24, 792, 2009.
18. Khan, S.A. and Rehman, S., On the use of Unified And-Or fuzzy aggregation operator for multi-criteria decision making in wind farm design process using wind turbines in 500 kW - 750 kW range, 2012, http://dx.doi.org/10.1109/fuzz-ieee.2012.6251306.
19. Nemes, C. and Munteanu, F., Optimal selection of wind turbine for a specific area, 2010, http://dx.doi.org/10.1109/optim.2010.5510577.
20. Perkin, S., Garrett, D., Jensson, P., Optimal wind turbine selection methodology: A case-study for Búrfell, Iceland. *Renew. Energy*, 75, 165, 2015.
21. Şağbanşua, L. and Balo, F., Multi-criteria decision making for 1.5 MW wind turbine selection. *Proc. Comput. Sci.*, 111, 413, 2017.
22. Dobakhshari, A.S. and Fotuhi-Firuzabad, M., A reliability model of large wind farms for power system adequacy studies. *IEEE Trans. Energy Convers.*, 24, 792, 2009.

23. Sarja, J. and Halonen, V., Case study of wind turbine sourcing: Manufacturer selection criteria, 2012, http://dx.doi.org/10.1109/epec.2012.6474953.
24. Chowdhury, S., Zhang, J., Messac, A., Castillo, L., Optimizing the arrangement and the selection of turbines for wind farms subject to varying wind conditions. *Renew. Energy*, 52, 273, 2013.
25. Chowdhury, S., Mehmani, A., Zhang, J., Messac, A., Market suitability and performance tradeoffs offered by commercialwind turbines across differingwind regimes. *Energies*, 9, 352, 2016.
26. Montoya, F.G., Manzano-Agugliaro, F., López-Márquez, S., Hernández-Escobedo, Q., Gil, C., Wind turbine selection for wind farm layout using multi-objective evolutionary algorithms. *Expert Syst. Appl.*, 41, 6585, 2014.
27. Bencherif, M., Optimum Selection of Wind Turbines. *Sci. J. Energy Eng.*, 2, 36, 2014.
28. Martin, K.A., Schmidt, M.F., Shelton, S.V., Stewart, S.W., Site specific optimization of rotor / generator sizing of wind turbines, 2007, http://dx.doi.org/10.1115/es2007-36144.
29. Jones, R.L., A concordance to biblical gothic. By Magnús Snædal. 2 volumes. Reykjavík: Institute of Linguistics, University of Iceland; University of Iceland Press, 1998. Vol. 1 = Introduction. Texts; pp. xxxiv, 70. Vol. 2 = concordance; pp. 1,257. Paper. $185.00. *J. Ger. Linguist.*, 13, 286, 2001.
30. Lyatkher, V., Collinear wind turbines (horizontal-axis wind turbines-HAWTs), 2013, http://dx.doi.org/10.1002/9781118720851.ch2.
31. Polat, O. and Tuncer, I.H., Aerodynamic shape optimization of wind turbine blades using a parallel genetic algorithm. *Proc. Eng.*, 61, 28, 2013.
32. Jowder, F.A.L., Wind power analysis and site matching of wind turbine generators in Kingdom of Bahrain. *Appl. Energy*, 86, 538, 2009.
33. Jureczko, M., Pawlak, M., Mężyk, A., Optimisation of wind turbine blades. *J. Mater. Process. Technol.*, 167, 463, 2005.
34. EL-Shimy, M., Optimal site matching of wind turbine generator: Case study of the Gulf of Suez region in Egypt. *Renew. Energy*, 35, 1870, 2010.
35. Saaty, R.W., The analytic hierarchy process-what it is and how it is used. *Math. Model.*, 9, 161, 1987.
36. Shirgholami, Z., Namdar Zangeneh, S., Bortolini, M., Decision system to support the practitioners in the wind farm design: A case study for Iran mainland. *Sustain. Energy Technol. Assess.*, 16, 1, 2016.
37. Dong, Y., Wang, J., Jiang, H., Shi, X., Intelligent optimized wind resource assessment and wind turbines selection in Huitengxile of Inner Mongolia, China. *Appl. Energy*, 109, 239, 2013.
38. Bagočius, V., Zavadskas, E.K., Turskis, Z., Multi-person selection of the best wind turbine based on the multi-criteria integrated additive-multiplicative utility function. *J. Civ. Eng. Manage.*, 20, 590, 2014.
39. Du, M., Yi, J., Mazidi, P., Cheng, L., Guo, J., A parameter selection method for wind turbine health management through SCADA data. *Energies*, 10, 253, 2017.

40. Khan, S.A. and Rehman, S., On the use of Unified And-Or fuzzy aggregation operator for multi-criteria decision making in wind farm design process using wind turbines in 500 kW - 750 kW range, 2012, http://dx.doi.org/10.1109/fuzz-ieee.2012.6251306.
41. Khan, S.A. (2016). Multi-criteria Decision-making for Selection of Wind Turbine with Rated Power of 1000 KW using Dubois and Prade Fuzzy Operator. *Int. J. Comput.,*, 04, 27-32.
42. Lee, A.H.I., Hung, M.C., Kang, H.Y., Pearn, W.L., A wind turbine evaluation model under a multi-criteria decision making environment. *Energy Convers. Manage.*, 64, 289, 2012.
43. Rehman, S. and Khan, S.A., Fuzzy logic based multi-criteria wind turbine selection strategy - A case study of Qassim, Saudi Arabia. *Energies*, 9, 872, 2016.
44. Bagočius, V., Zavadskas, E.K., Turskis, Z., Multi-person selection of the best wind turbine based on the multi-criteria integrated additive-multiplicative utility function. J. *Civ. Eng. Manage.*, 20, 590, 2014.
45. Rehman, S. and A., S., Multi-criteria wind turbine selection using weighted sum approach. *Int. J. Adv. Comput. Sci. Appl.*, 8, 2017.
46. Rehman, S. and Khan, S.A., Goal programming-based two-tier multi-criteria decision-making approach for wind turbine selection. *Appl. Artif. Intell.*, 33, 27, 2019.
47. Sánchez-Lozano, J.M., Teruel-Solano, J., Soto-Elvira, P.L., Socorro García-Cascales, M., Geographical Information Systems (GIS) and Multi-Criteria Decision Making (MCDM) methods for the evaluation of solar farms locations: Case study in south-eastern Spain. *Renew. Sustain. Energy Rev.*, 24, 544, 2013.
48. Basak, I. and Saaty, T., Group decision making using the analytic hierarchy process. *Math. Comput. Model.*, 17, 101, 1993.
49. Saaty, T.L. and Kearns, K.P., *Systems*, 1985, http://dx.doi.org/10.1016/b978-0-08-032599-6.50007-6.
50. MacCormac, E.R., Review of: "Decision making for leaders: The analytical hierarchy process for decisions in a complex world" by Thomas L. Saaty, Belmont, Calif.: Lifetime Learning Publications, 1982, 291 pages, $24.95. *Eng. Econ.*, 29, 74, 1983.
51. Labib, A., *Introduction to the analytic hierarchy process*, 2014, http://dx.doi.org/10.1007/978-3-319-12502-2.
52. Emrouznejad, A. and Ho, W., Analytic hierarchy process and fuzzy set theory, 2017, http://dx.doi.org/10.1201/9781315369884-1.
53. Malczewski, J., GIS-based multicriteria decision analysis: A survey of the literature. *Int. J. Geogr. Inf. Sci.*, 20, 703, 2006.
54. Saaty, T.L., How to make a decision: The analytic hierarchy process. *Eur. J. Oper. Res.*, 48, 9, 1990.

55. Myers, M.L., *Guidelines for estimating nonfuel operation and maintenance costs for alternative nuclear power plants*, Office of Scientific and Technical Information (OSTI), 1979.
56. 1. UNESCO, *Taking a whole of government approach to skills development*, 2018.
57. Doustmohammadi, N. and Babazadeh, R., Design of closed loop supply chain of wood plastic composite (WPC) industry. *J. Environ. Inform.*, 35, 94, 2020.
58. Basir, A., Siraj, A.S., Huq, K.R.M., Analysis of building energy efficiency and integration of renewable energy sources, 2015, http://dx.doi.org/10.1109/iceeict.2015.7307531.
59. Special report on solar PV global supply chains, 2022, http://dx.doi.org/10.1787/9e8b0121-en.
60. Marszal, A.J., Heiselberg, P., Bourrelle, J.S., Musall, E., Voss, K., Sartori, I., Napolitano, A., Zero Energy Building - A review of definitions and calculation methodologies. *Energy Build.*, 43, 971, 2011.
61. Gass, S. and Saaty, T., The computational algorithm for the parametric objective function. *Nav. Res. Logist. Q.*, 2, 39, 1955.

9

Effect of Aerodynamics on Wind Turbine Design

Mahadi Hasan Masud[1*], Md. Forhad Hossain Hemal[1], Mim Mashrur Ahmed[1], Md. Fyruz Ibna Alam Taki[1], Md. Hasibul Hasan Himel[1,2], Anan Ashrabi Ananno[3] and Peter Dabnichki[4]

[1]*Department of Mechanical Engineering, Rajshahi University of Engineering and Technology, Rajshahi, Bangladesh*
[2]*Department of Mechanical and Production Engineering, Ahsanullah University of Science and Technology, Dhaka, Bangladesh*
[3]*Division of Product Realization, Department of Management and Engineering, Linköping University, Linköping, Sweden*
[4]*School of Engineering, RMIT University, Bundoora Campus, Melbourne, VIC, Australia*

Abstract

The huge potential of wind energy to generate power has made the study of wind turbines very exciting. Wind turbines should be designed in such a way that they can offer the best reliability along with maximum efficiency. As the performance of wind turbines relies on different aerodynamic properties, this chapter aims to illustrate the key aerodynamic properties that must be taken into consideration during turbine design for improved performance. This study starts with explaining the effect of different air properties like density, viscosity, temperature, pressure, and humidity on the turbine performance. Finally, aerodynamic properties that define the turbine performance, such as solidity, blade numbers, pitch, pitch angle, strut effects, height-to-radius ratio, and Reynolds number, are discussed in detail. It is concluded that all these properties are equally important and are to be given proper consideration during the process of wind turbine design to achieve maximum efficiency with satisfactory reliability.

Keywords: Wind turbine, aerodynamics, reliability, momentum theory, Reynolds number

**Corresponding author*: masud.08ruet@gmail.com; masud.08@me.ruet.ac.bd

Inamuddin, Tariq Altalhi and Mohammad Luqman (eds.) Wind Energy Storage and Conversion: From Basics to Utilities, (207–246) © 2024 Scrivener Publishing LLC

9.1 Introduction

Global electricity demand is increasing, and current over-reliance on fossil-based energy technology is negatively impacting the environment. Therefore, extensive research is required to harness electricity from renewable sources. Though there are different options for renewable sources, wind energy, due to its satisfactory efficiency along with availability, is widely utilized globally [1, 2]. The Global Wind Energy Council statistics [3] indicate that, in 2021, 837 GW of power was harnessed from wind energy [3]. The planned installation of 6,044 GW of wind energy plants by 2050 has the potential to meet over one-third of the global electricity demand. Such action will reduce up to 6.3 GTons of CO_2 emissions [4]. The basic projections outlined above demonstrate the substantial impact wind energy can have in mitigating climate change and transitioning towards a more sustainable future. Moreover, the advancement of wind turbine technology and management has been closely linked to the extensive adoption of wind power throughout the past two decades [5].

Wind turbine development has been geared up after the research done by Joukowsky and Betz [6, 7]. Glauert's [8] development of the blade-element momentum (BEM) theory opened a new window for designing the turbine rotors. These advancements have allowed modern horizontal-axis wind turbines (HAWTs) to achieve maximum power coefficients close to the theoretical limit [9]. Tian *et al.* [10] developed and examined an improved vertical-axis wind turbine (VAWT) having overlapped savonius rotors positioned next to each other. With a tip speed ratio (TSR) of 0.4, their research found a considerable gain in efficiency with a 46.95% increase. Another research of savonius type wind turbine has been carried out by Kaya and Acir [11] based on the Taguchi optimization method and concluded that the output power coefficient increased by approximately 17.9% when the TSR was set to 0.8. However, it is very challenging to make a precise prediction about wind turbine performance under natural conditions due to presence of boundary layer across the turbine blade [12]. Research conducted by Costa Rocha *et al.* [13] concluded that a suitable blade pitch angle controller can optimize wind turbine performance. To reduce the effect of axial loads, Lacker [14] introduced the PID controller to maintain the nominal conditions of the rotor. Aerodynamic imbalance is another concerning parameter for the fatigue damage of the wind turbine structure. Extensive analyses by Castellani *et al.* [15] concluded that wake generation is one of the main reasons for this imbalance, which is responsible for poor power output of wind turbine. Madsen *et al.* [16] utilized computational fluid dynamics to ensure proper optimization of the blade tip shape, while Du *et al.* [17] presented a deep learning method to predict the output aerodynamic performance of a wind turbine blade. They also utilized neural networks capable of generating smooth and comprehensive blade profiles. Chetan *et al.* [18] investigated the aerodynamic challenges posed by overlong blades. Duthe *et al.* [19] investigated a critical issue that impacts the dependability of wind turbines, i.e., the erosion of rotor blades occurring on the different edge point. Such erosion diminishes performance and causes imbalances due to the load variation, amplified noise pollution. Thus, maintenance requires more expenses. If neglected, this problem can eventually lead to a failure in the functionality of the blades.

The mechanical energy generated by wind turbines is transformed into electricity. The turbine performance depends on the wind velocity, rotor area and density of the air [20]. However, the actual power output is less than the theoretical one due to wind shear stress responsible for considerable fatigue load on the turbine blade [22]. Variations of external climate conditions (temperature, humidity, and density of air), the effect of wakes around the rotor blade [21], yak angle [22], angle of attack (AOA) and TSR are the most critical aerodynamic variables that affect the power output. Moreover, to ensure proper power optimization, it is necessary to reduce axial loads acting on the turbine blades [23]. Proper optimization of those aerodynamic variables can ensure maximum power output. Wind turbine aerodynamics represents a prominent area of study within contemporary aerodynamics. The study of wind turbine aerodynamics involves discussing various methods for modeling and estimating the aerodynamic forces exerted on a wind turbine's solid components, explicitly focusing on the rotor blades. Aerodynamics plays a pivotal role in predicting the wind turbine performance and is an essential requirement for their design, advancement, and optimization [24].

Various researchers studied different parameters [25–28] and showed their effects on wind turbines performance, but a critical study exploring all the aerodynamic variables is still missing. This study provides a detailed analysis of how various aerodynamic variables impact the turbine performance, the interrelationships between the variables and effects on power generation, and recommendation has been given for ensuring optimum wind turbine performance. In this chapter, the authors will embark on a journey to uncover the fundamental principles of aerodynamics that shape the design and operation of wind turbines. An exploration of the different key elements of a wind turbine including blades, tower and nacelle and a discussion of their specific aerodynamic roles is presented. This chapter focuses on assessing the impact of different parameters like, wind velocity and blade design on the turbine performance. This chapter briefly discusses the effects of air properties on wind turbines and classical momentum theory, which is popular for being one of the most fundamental theories for understanding wind turbine aerodynamics, followed by a detailed explanation of the interrelationship between wind turbine and aerodynamics.

9.2 Air Properties Affecting Wind Turbines

Understanding interrelationship between turbines and aerodynamics seeks an in-depth understanding of air properties that directly or indirectly affect wind turbine performance. Wind kinetic energy is transformed into mechanical energy by wind turbines, which are then utilized to generate electricity [29]. Therefore, suitable air properties are required for optimum performance of the wind turbine. For better performance, the shape of the blade and the associated angles are carefully designed to reduce the drag force while increasing the lift force [30]. Apart from shape and angle of attack, several factors influence the drag and lift forces, such as blade design and air properties. Air properties like wind speed, wind direction, air density, turbulence, viscosity, humidity, temperature, and pressure affect the aerodynamic performance of turbines. Wind speeds should be optimum for the optimum wind turbine performance as excessive high speed may damage the turbine blades, and low speed cannot provide enough thrust for the rotation of the blade. For higher power output, the wind should hit the turbine blade orthogonally all the time. If the air density is low, it will result in less mass flow through the turbine resulting in an overall reduction in the power output. Moreover, increased viscosity, turbulent air and higher humidity can limit the turbine performance [31–34].

Air density has significant impact on the turbine aerodynamics. The turbine performance improves when it is exposed to high-density airflow passing through its rotor. Moreover, lift force increases as a result of the high-density airflow. Therefore, a high-density air flow ensures high turbine efficiency, allowing more energy extraction. However, the longer and broader blades can compensate for lower lift forces in low-density locations. Blade design needs to be modified for various air density circumstances. The analysis of the turbine performance and energy production requires an accurate understanding of air density. Therefore, the design of wind turbines for ensuring maximum efficiency and power generation is directly connected to the air density [31].

Viscosity significantly impacts the aerodynamics of wind turbines. Viscosity refers to the resistance of a fluid to flow that affects the behavior of the air when it flows around the blades of the wind turbine. At minimum wind speeds, air viscosity can provoke it to stick with the turbine blade surface and creates a boundary layer. This boundary layer increases the drag on the blades and reduces the turbine efficiency. At higher wind speeds, the viscosity of air can cause turbulence and separation of the airflow over the blades, leading to reduced lift and increased drag, thus resulting in poor turbine performance [35].

Temperature is a crucial factor that significantly impacts the turbine performance as the air density changes with temperature. When the temperature increases, the density decreases, resulting in less mass flow around the turbine and causing less power output. Moreover, temperature differences in the atmosphere can create areas of high and low pressure that alter the wind direction [32].

Air pressure directly affects the lift and drag force on the blades. Flowing the wind over the blade curved surface, it creates a zone of low pressure along with high pressure respectively on the upper surface and lower surface. The pressure variance between the two produces lift by generating a force that elevates the blades. The amount of lift produced is proportional to the pressure variance between the upward and downward surfaces of the blade. The pressure gradient across the blade generates drag force that can be minimized by appropriate design [33].

Air humidity directly affects the air density, which, in turn, impacts the wind turbine performance. High humidity results in less air density, reducing

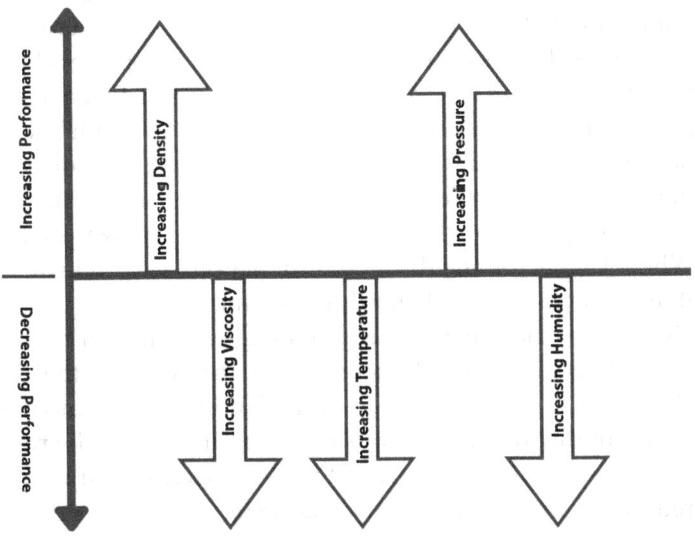

Figure 9.1 Air properties affecting wind turbine performance.

the effectiveness of the turbine as lower density is responsible for the lower air mass flow through the turbine. Consequently, a wind turbine generates less power in the humid condition than the dry condition. Humidity also affects the airflow over the surface of the blades as high humidity decreases air velocity [34].

Figure 9.1 demonstrates that air properties significantly affect the wind turbines' performance. In short, high viscosity, temperature, and humidity reduce the turbine performance. However, high pressure and density are suitable for improved power output of the turbine.

9.3 Classical Blade Element Momentum Theory

The very first developed classical momentum theory by Rankine and Froude for one-dimensional flow which was developed by Glauert to understand two-dimensional flow integrating spinning motion in the wake is essential to analyze the aerodynamic effects on turbines [38]. It is a mathematical approach by which the aerodynamics of a wind turbine rotor blade can be examined. The wind turbine blade performance can be assessed by the analysis of the individual component of the overall system. The significance of the BEM theory lies in its capacity to offer a straightforward yet precise technique for predicting wind turbine performance. Hence, for the assessment of aerodynamic loads and power output, this theory is repeatedly for a wind turbine [39] while also being computationally efficient in predicting wind turbine performance [40]. The approach of the BEM theory is founded upon the approach of incompressible, steady, and inviscid flow that has no circumferential and radial dependency which can be obtained from the conversion of mass and momentum law [41]. The main assumption of BEM is that it disregards the wake rotation effect [42]. However, the wake rotation effect is significant when analyzing propellers as it increases the power coefficient of the rotor having low tip speed ratios (TSR). When TSR <2, at the disc, axial induction factor remains smaller than half of that at the infinitely far downstream.

Negative flow in the wake happens at a higher induction factor (≥0.5), which is an issue in the BEM theory [38]. This violates the Rankine–Froude flow theory, which is a limitation of BEM theory. Another shortcoming arises in that the theory considers the flow within the rotor blade as a two-dimensional component, and the airfoil section characteristics are considered to be constant across the blade span. This ignores the effects of three-dimensional flow phenomena, like wake turbulence and vortex shedding, which can significantly affect the rotor performance [42]. Again, the

BEM theory assumes the flow through the rotor blades to be consistent and stable, ignoring the impacts of unsteady flow phenomena like a dynamic stall, which can result in significant differences between the expected and the actual rotor performance [38].

The power production generally becomes higher when the induction factor increases, but it also increases the structural loads on the blades. Moreover, extremely high induction factors result in power output reductions because of the increase in turbulent wakes [43]. The induction factor must be optimized to maximize power output while minimizing blade loads. Figure 9.2 illustrates the rotor's wake and upstream induction zones of a wind turbine.

Ignoring the relatively small tangential induction factor (swirl), the rate of mass flow ($d\dot{m}$) over the disk annular portion is:

$$d\dot{m} = \rho dA(V_\infty - v_i) = 2\pi\rho(V_\infty - v_i)y\ dy \tag{9.1}$$

Here V_∞ and v_i represent velocity at free stream and induced velocity in axial direction in ms^{-1}, respectively, and ρ represents the density of air.

The incremental thrust on the annulus in general form (dT) and in coefficient form (dC_T) can be calculated using Equations 9.2 and 9.3, respectively.

$$dT = 2\rho(V_\infty - v_i)v_i dA = 4\pi\rho(V_\infty - v_i)v_i y\ dy \tag{9.2}$$

$$dC_T = \frac{dT}{\frac{1}{2}\rho A V_\infty} = 8\left(1 - \frac{v_i}{V_\infty}\right)\frac{v_i}{V_\infty}\left(\frac{y}{R}\right)d\left(\frac{y}{R}\right) \tag{9.3}$$

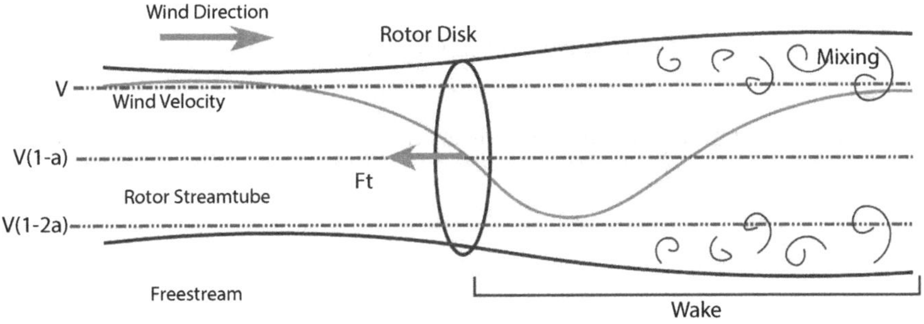

Figure 9.2 Wind turbine rotor's wake and upstream induction zones.

Here R represents the rotor radius in meters. Taking dimensionless blade radius, $r = y/R$, Equation 9.3 becomes:

$$dC_T = 8\left(1 - \frac{V_i}{V_\infty}\right)\frac{V_i}{V_\infty} rdr = 8(1-a)ardr \tag{9.4}$$

where induction factor $a = \frac{V_i}{V_\infty}$. The thrust coefficient may be expressed using the blade element method as follows:

$$dC_T = \sigma(X_{TSR})^2 C_l r^2 dr \tag{9.5}$$

where X_{TSR} is the TSR and σ is the rotor solidity. Equating Equations 9.4 and 9.5, the following expression can be obtained:

$$8(1-a)ar\ dr = \sigma(X_{TSR})^2 C_l r^2 dr \tag{9.6}$$

The coefficient (C_l) and the inflow angle (φ) can be determined by the following equations:

$$C_l = C_{l\alpha}(\theta + \varphi) \tag{9.7}$$

$$\varphi = \frac{1-a}{rX_{TSR}} \tag{9.8}$$

Here θ is the blade pitch angle and $C_{l\alpha}$ is the sectional lift slope with the angle of attack. Thus, Equation 9.6 becomes:

$$8(1-a) = \sigma X_{TSR} C_{l\alpha}(X_{TSR}\theta r + (1+a)) \tag{9.9}$$

This subsequent equation, with some adjustment, can be stated as follows:

$$a^2 - \left(\frac{\sigma X_{TSR} C_{l\alpha}}{8}\right)a + \frac{\sigma\ X_{TSR} C_{l\alpha}(X_{TSR}\ \theta\ r + 1)}{8} = 0 \tag{9.10}$$

$$a(r, X_{TSR}) = \left(\frac{\sigma X_{TSR} C_{l\alpha}}{16} + \frac{1}{2}\right) - \sqrt{\left(\frac{\sigma X_{TSR} C_{l\alpha}}{16} + \frac{1}{2}\right) - \frac{\sigma X_{TSR} C_{l\alpha}(X_{TSR}\ \theta\, r + 1)}{8}} \tag{9.11}$$

Equation 9.11 is considered the principal equation of BEM theory. Momentum theory is only valid for the induction factor from 0 to 0.50 since beyond that the turbine gradually goes to the TWS and VRS.

Therefore, turbine performance can be assessed by induction factor "a" if the sufficient data on blade element parameters, pitch angle, rotational speed, and air density are available. Moreover, the load can be analyzed by estimating the loads experienced by various wind turbine components.

9.4 Aerodynamic Performance Testing

The turbine performance can be enhanced by capturing the maximum possible wind energy and converting it into usable power [36, 37]. In this section, the methods of aerodynamic performance testing are discussed, along with their importance.

There are two main methods of aerodynamic performance testing: (1) wind tunnel testing and (2) field testing. The turbine blade or model is placed in a controlled airflow during wind tunnel testing, and aerodynamic forces and moments over the blades are measured [38]. The performance evaluation of various types of blades is possible by this method because it allows for the precise control of some testing parameters like the wind speed and AOA [39]. Field testing is used for the assessment of the aerodynamic performance in outside conditions [40]. However, field testing is subjected to environmental conditions that can affect the accuracy and reliability of the data. Both methods have both positive and negative aspects, and the method selection depends on the types of requirements.

9.4.1 Wind Tunnel Testing and Field Testing

Wind tunnel testing is essential to evaluate the efficiency of the wind turbines in a predictable and controlled environment. The wind tunnel is made up of three parts such as a test section where the model is placed for observation, where the wind is controlled in the real world, a settling chamber where the wind is conditioned, and a wind-generating and flow apparatus.

The wind speed and other important parameters are measured by a variety of sensors, and the performance of wind turbines such as torque, output power, and rotational speed are collected using data acquisition systems.

Wind tunnel testing is more reliable when there is regulated flow, although the obstruction effects often affect the reliability. Because of the unsteady, asymmetric, separated, and highly turbulent flow field that surrounds them, VAWTs create a special aerodynamic interruption in wind tunnels [41]. The literature suggests correcting the output of the VAWT if the blockage ratio exceeds 5% [50].

A digital manometer coupled with pitot-static tube is used to measure V_∞. The free-stream flow's turbulence intensity (Tu) can be calculated by the following equation [41]:

$$\mathrm{Tu}(\%) = \frac{v_{rms}}{\bar{V}} \times 100 \tag{9.12}$$

where v_{rms} and $\bar{V}$ represent root-mean-squared velocity fluctuations at the streamwise direction and time-average streamwise velocity of wind, respectively. The root-mean-squared velocity is calculated as:

$$v_{rms} = \sqrt{\sum_{i=1}^{n} \frac{\left(v_i - \bar{V}\right)^2}{n-1}} \tag{9.13}$$

where v_i represents the induced axial velocity and n represents the number of total observations. Turbulence intensity or Tu is repeatedly used in the testing of wind tunnels to describe the degree of turbulence in the test section and to observe the performance of the model being tested due to turbulence. By measuring the turbulence strength in a wind tunnel, researchers can determine the turbulence in the test section and compare the results of numerous experiments carried out under similar conditions. External loads are connected to each turbine outlet for the determination of power output. The amount of power consumed (P_l) by the resistor is then determined using the circuit's effective value of alternating current, I as shown in Equation 9.14:

$$P_l = R_l I^2 \tag{9.14}$$

where R_l represents the electrical resistance of the load [49]. Wind tunnel testing has several advantages over field testing. From wind tunnel testing, the computational results are usually validated [42]. Researchers can increase the precision of their models and accurately predict the turbine performance in actual operating conditions by comparing the outcomes of wind tunnel testing with those predicted by computational models. Wind turbines are frequently installed in turbulence-prone areas, which can significantly affect their performance [43]. The wind turbine design can be improved with the help of this information to make them more resistant to turbulence. Different wind turbine designs can be tested and analyzed in a controlled environment using wind tunnel testing [44]. Field testing is a vital tool for evaluating and improving wind turbine performance. The outcomes of this testing can be used to increase the overall effectiveness and dependability of wind energy systems, which will eventually assist in the development of the renewable energy sector.

9.4.2 Performance Testing of a Counter-Rotating Wind Turbine System

A counter-rotating turbine system (C/R system) [45] utilizes patented bevel-planetary gear arrangements to combine the VAWT with the conventional HAWT systems as shown in Figure 9.3. The inner 30% area of the blade, also called dead zone, faces a low sweeping speed [45]. To cover up the dead zone, an auxiliary rotor, whose diameter is half of that of the main rotor, is used to generate additional torque in the main rotor area. This system has significant benefits, especially in terms of aerodynamics. Moreover, turbine performance can be enhanced using an auxiliary rotor of a smaller dimension. The weight of the nacelle is greatly reduced by the generating system's placement in the non-rotating section, enabling a free-yaw feature. To analyze this system, it is assumed that the primary and the auxiliary rotors do not interfere with each other aerodynamically.

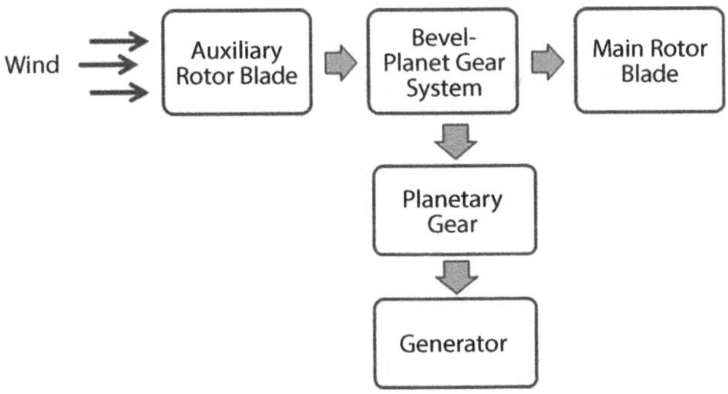

Figure 9.3 Schematic diagram of a counter-rotating (C/R) wind turbine system.

Stream tube geometry is depicted in Figure 9.4 [45]. Let V, V_1, and V_o be flow components towards the axis of stream tube and ρ and A be the air density and disk area, respectively. The air stream releases the power (P) which can be expressed as follows:

$$P = \frac{1}{2}\rho AV(V_o + V_1)(V_o - V_1) \tag{9.15}$$

The power coefficient (C_p), as mentioned earlier, can be represented as:

$$C_p = \frac{P}{0.5\rho V_o^{\,3} A} = 4a(1-a)^2 \tag{9.16}$$

Based on Glauert's actuator disk concept, the BEM theory is used to obtain the torque and thrust of the rotor [45]. A typical blade section aerodynamics is illustrated in Figure 9.5. The blade's geometric pitch is specified as follows:

$$\theta = \varphi - \alpha \tag{9.17}$$

where θ and φ represent the induced flow angle and angle of attack, respectively.

Considering that a′ is the rotational induction, the following expression can be used to express the induced angle:

$$\varphi = \tan^{-1}\left(\frac{1-a}{\lambda_r(1+a')}\right) \tag{9.18}$$

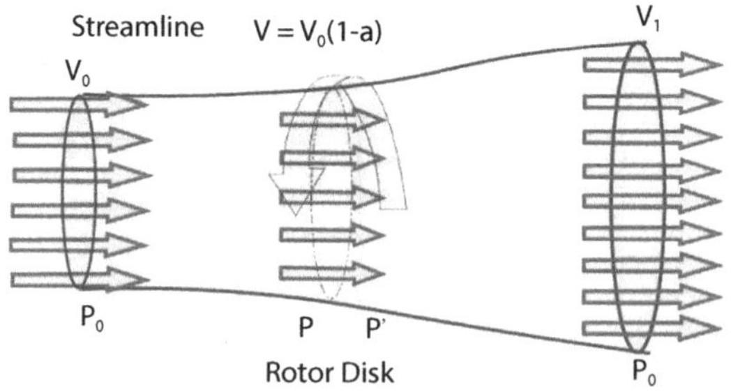

Figure 9.4 Geometry of the stream tube.

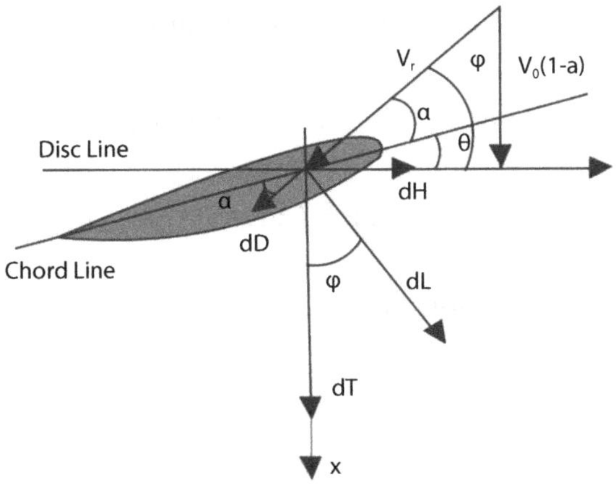

Figure 9.5 Blade section aerodynamics.

where $\lambda_r = \dfrac{\Omega r}{V_o}$ and $a' = \dfrac{1-3a}{4a-1}$

where λ_r and Ω represent induced flow ratio and rotor speed in RPM, respectively. Inflow ratios and induction factors can be used to express the relative air velocity (V_r) as follows:

$$V_r = V_o\sqrt{\{(1-a)^2 + \lambda_r^2(1-a)^2\}} \tag{9.19}$$

Aerodynamic forces operating over a short blade section, dr, can be estimated as follows:

$$dL = \frac{1}{2}\rho V_r^2 c \, dr \, c_l \tag{9.20}$$

$$dD = \frac{1}{2}\rho V_r^2 c \, dr \, c_d \tag{9.21}$$

where infinitesimal drag and lift forces are represented by dD and dL, respectively, and c represents chord length. Moreover, c_d and c_l represent the drag and lift coefficients, respectively. Additionally, the infinitesimal thrust (dT), power generated (dP), and developed torque (dQ) are calculated as follows:

$$dT = dL\cos\varphi + dD\sin\varphi \tag{9.22}$$

$$dQ = (dL\sin\varphi - dD\cos\varphi)r \tag{9.23}$$

$$dP = \Omega dQ \tag{9.24}$$

Combining contributions from small disk portions, the equations become as follows:

$$T = \frac{N_b}{2}\int_0^F \rho c V_r^2 (c_l \cos\varphi + c_d \sin\varphi) dr \tag{9.25}$$

$$Q = \frac{N_b}{2}\int_0^F \rho c V_r^2 (c_l + \sin\varphi - c_d \cos\varphi) r\, dr \tag{9.26}$$

$$P = \Omega Q \tag{9.27}$$

where T is the total thrust, and Q and P are the total torque and power, respectively. These power, torque, and thrust estimation are crucial for performance analysis. Some previous performance studies are presented in Table 9.1.

Numerous works have been performed recently to enhance the turbine performance. As observed from Table 9.1, focus has been given to maximizing the power coefficient of the turbine to ensure high efficiency. A maximum power coefficient of 0.4288 was achieved during a wind tunnel test done by Talavera and Shu [52].

9.5 Effect of Aerodynamics on Wind Turbine Design Parameters

A number of aerodynamic factors directly affect the wind turbine design, such as solidity, number of blades, different ratio, pitch angle, Reynolds number, and strut effect. The effects of these factors are summarized in the following discussions.

Table 9.1 Summary of the performance testing of wind turbines in recent years.

	Specifications of the test				Input parameters					Output parameters			
Location	Test type	Dimensions	Airfoil profile	No. of blades	TSR, λ	Pitch angle, β (°)	Incoming wind velocity (m/s)	Reynolds number, Re	Turbulent intensity, Tu (%)	Maximum power co-efficient, Cp	Tip speed ratio, λ	Comment	Ref.
Mie University, JAPAN	Wind tunnel test	Rotor diameter = 2 m	NACA 0021	2	0.454–2.761	4°–8°			Less than 0.5	0.184	2.193 at β = 6°	Highest wind velocity was taken 30 ms^{-1} with a rotational range of 0–6,000 rpm	[46]
		Blade height = 1.2 m											
		Chord length = 0.265 m											
		Hub height = 5 m			0.592–2.914			2.16 × 10^5–2.89 × 10^5		0.175	2.201 at Re = 2.53 × 10^5		
		Wind tunnel outlet diameter = 3.6 m											
		Wind tunnel length = 4.5 m											
	Field test	Hub height = 5 m			0.798–3.183	6°	3 -11			0.209	2.211 at wind velocity of 6 to 7 m/s	Direction of the inflow velocity was assumed to be 90°	
					0.925–3.175				0.25–0.40	0.1772	2.237 at Tu = 0.35		
		Hub height = 0.2 m											

(Continued)

Table 9.1 Summary of the performance testing of wind turbines in recent years. (*Continued*)

Location	Specifications of the test				Input parameters					Output parameters		Comment	Ref.
	Test type	Dimensions	Airfoil profile	No. of blades	TSR, λ	Pitch angle, β (°)	Incoming wind velocity (m/s)	Reynolds number, Re	Turbulent intensity, Tu (%)	Maximum power co-efficient, Cp	Tip speed ratio, λ		
WiRE laboratory of EPFL	Wind tunnel test	Diameter of the wind turbine = 15 cm	NACA 0012	3	0.492–6.622		Less than 10 m/s	Less than 4 × 10^4		0.395	3.135 at 8 m/s free stream velocity	Minimum blockage effect is observed	[47]
		Wind tunnel is 28 m (L) × 2.6 m (W) × 2 m (H)											
		Free stream velocity ranged between 4 and 8 m/s											
N/A	Wind tunnel test	Wind tunnel is 14.6 m (L) × 1.2 m (W) × 1.2 m (H)		2	4.304–7.902		13	Laminar	Less than 0.5	0.1516	5.390 at diameter of 15		[48]
		Blade diameter = 8 in			4.165–9.225			Turbulent		0.4288	7.695 at diameter of 10		
		Blade diameter ranges from 5 to 15											

9.5.1 Solidity

VAWT solidity is the ratio of turbine swept area to the blade planform area as depicted in Figure 9.6 [49]. Wind turbine performance is significantly influenced by the blades' solidity. Higher solidity results in broader and thicker blades that can enhance the rotor's torque and increase the turbine's drag. Lower solidity indicates thinner and narrower blades that can minimize drag and enhance the effectiveness of the turbine [50]. Several studies have defined this solidity value as either NcD or NcR. Figure 9.7 depicts the dependence of power coefficient on TSR for different solidity values.

Therefore, it is crucial to choose a solidity that optimizes effectiveness without requiring too much blade material. As observed from Figure 9.7, the operating limit of the VAWT decreases when solidity increases [50]. From a power generation standpoint, a high-solidity turbine is not preferred due to its unfavorable operational characteristics. When a turbine rotates slowly and possesses high torque, it will experience dynamic stalls more frequently [51]. Conversely, very low solidities are undesirable since they result in low peak efficiencies and large rotational velocities [52].

It is clear that increasing turbine solidity reduces turbine output by increasing the interactions between the blades and wakes. Hence, it is crucial to find a balance between solidity, TSR, and the impact of flow blockage in order to optimize the turbine efficiency. By altering the length of the blade chord and the number of blades, the solidity can be modified while keeping all the other factors constant. It is suggested to keep the solidity between 0.2 to 0.3 to achieve better aerodynamic efficiency.

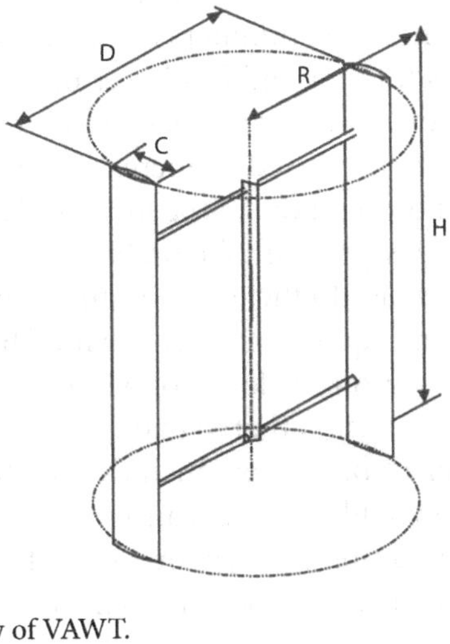

Figure 9.6 Isometric view of VAWT.

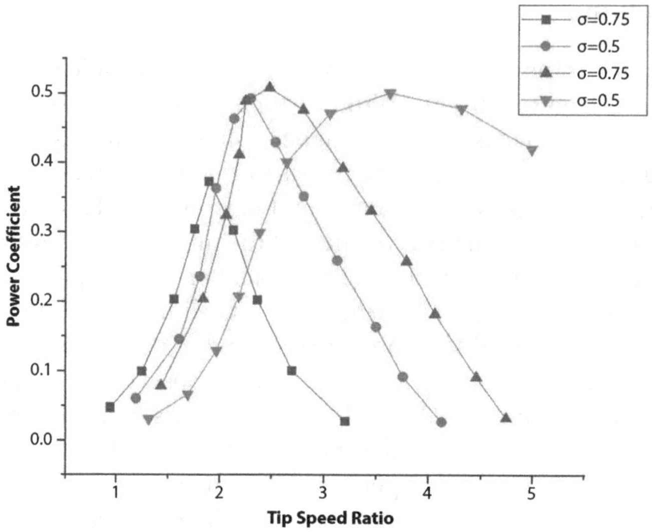

Figure 9.7 Impact of solidity on power coefficient (here σ = solidity of wind turbine) [50].

This comparatively low solidity may increase the efficiency and lower the material costs for bulky turbines [51]. TSR between 3 to 4 are more preferable in practical cases to eliminate the possibility of dynamic stall [53].

9.5.2 Number of Blades

A practical design decision is the number of blades, which is eventually a balance between the blade stiffness, effective aerodynamics, and economic factors [54]. As the blade numbers (for a given solidity of a turbine) increases, the blade gets thinner and has lower bending resistance. From a structural design point of view, these blades would require more support or struts. As a result, the drag produced by increased struts can negate the performance improvement brought on by the increased blade numbers. While the quantity of blades is a significant factor, having more than two or three blades can result in decreased efficiency. A study by Blackwell *et al.* [55] highlighted that to attain equivalent performance between the two- and three-bladed turbines, rotational velocity or wind speed has to be increased. The analysis led to the conclusion that to maintain a high Reynolds number, minimizing the number of blades as much as possible is recommendable [55]. Augmenting the Reynolds number can also enhance the turbine's aerodynamic performance and prevent dynamic stall at higher AOA [56]. The influence of blade number on the power coefficient for a definite solidity is depicted in Figure 9.8. From two to three blades, the performance of the rotor changes noticeably. However, more than three blades do not considerably enhance the rotor performance and are usually neglected [57].

Figure 9.9 illustrates the variation of power coefficients with TSR for two- to four-bladed conditions with similar solidities. It is confirmed that

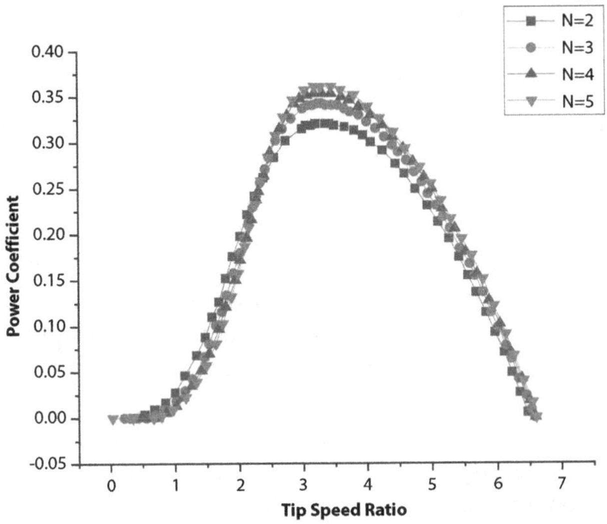

Figure 9.8 Variation of power coefficient with TSR for various blade numbers [57].

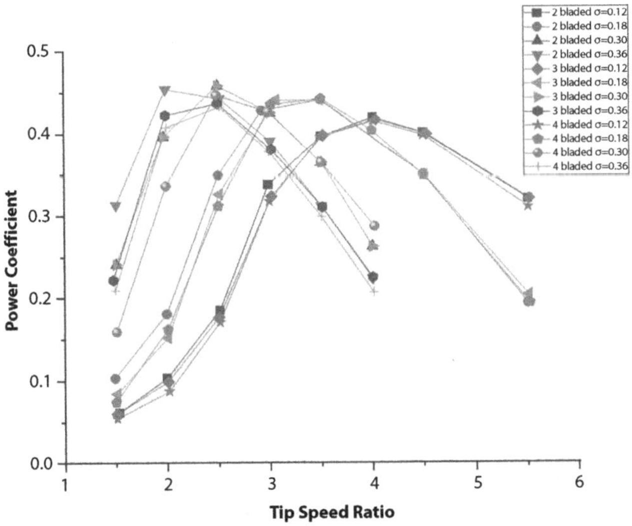

Figure 9.9 Power coefficient curve of two to four bladed VAWTs [58].

the coefficient of power does not depend on the blade numbers. With minimum solidity and constant speed, urban VAWTs often run at medium to high speeds with constant Reynolds number, where the C_p is mostly independent on blade numbers. Urban VAWTs, as previously indicated, must have a low solidity for improved performance [58].

9.5.3 Different Ratios

9.5.3.1 Chord/Radius Ratio (c/R)

The c/R ratio can be defined as the ratio of the chord length to the rotor radius [59]. With a lower c/R ratio, energy extraction from the blades is increased, as well as the aerodynamic drag. Although these turbine blades need higher cut-in speeds, they can start operating at lower wind speeds. Blades with a low c/R ratio provide higher torque and power output but need more complex control systems. The blades are thicker, which increases the structural strength but results in more weight. Although a high c/R ratio increases noise and turbulence, it ensures performance and stability enhancement due to less lift and drag. This c/R ratio is optimized by designers considering the influence on the environment, power production, structural integrity, and aerodynamic efficiency [60].

As per the data presented in Figure 9.10, it can be observed that there is a positive correlation between the power coefficient and chord length.

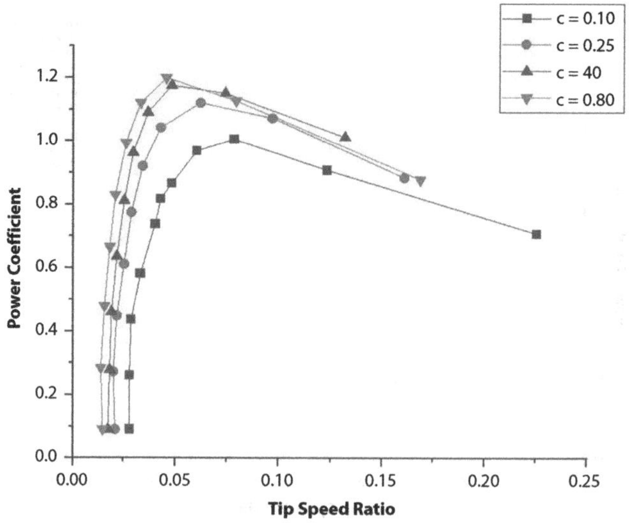

Figure 9.10 Effect of chord length on wind turbine performance [61].

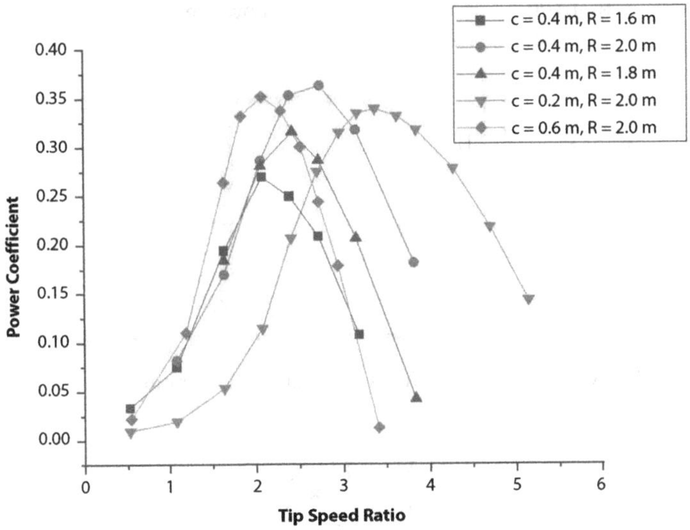

Figure 9.11 Relationship between turbine performance and c/R ratio [61].

Enhancing the chord from 0.10 to 0.80, the maximum power coefficient increases. Additionally, there is a slight increase in the stall angle as the chord length increases. Figure 9.11 visually demonstrates the relationship between c/R and turbine performance. It is not advisable to solely rely on increasing the radius to increase the turbine performance. Doing so would escalate capital costs and associated operational complexities. The decision on the c/R ratio will subsequently affect the turbine performance, as stated in [61].

The power coefficient exhibits a nearly consistent value for identical chord lengths at lower TSRs, as observed in Figure 9.11. By increasing the radius, the c/R ratio decreases, resulting in a decrease in dynamic stall. Moreover, turbine performance slightly increases with "c" as it will eventually increase the Reynolds number, thereby enhancing the aerodynamic performance of an airfoil. However, an extended chord length would also result in a significant dynamic stall. Describing the effect of c/R ratio on the power coefficient is complex due to the intricate interaction between the c/R ratio and the operating range of TSR. Specifically, as the chord–radius ratio decreases, the operating range of TSRs expands [61].

9.5.3.2 Height-to-Radius Ratio (H/R)

This ratio is obtained by multiplying the blade aspect and chord-to-radius ratio denoted as H/c and c/R:

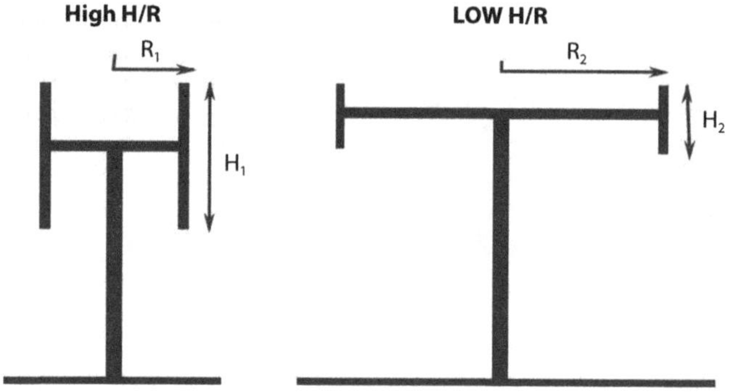

Figure 9.12 The swept area of two VAWTs with varying H/R ratios [50].

$$\frac{H}{R}=\frac{H}{c}\times\frac{c}{R} \tag{9.28}$$

Elevating the turbine height has the potential to mitigate the effects of turbulence at ground level, resulting in a more consistent and stable wind stream [62]. Higher wind speed can be achieved for smoother surfaces and places where the obstruction is comparatively lower. Figure 9.12 illustrates the swept area of two VAWTs with different H/R ratios. It is preferable to

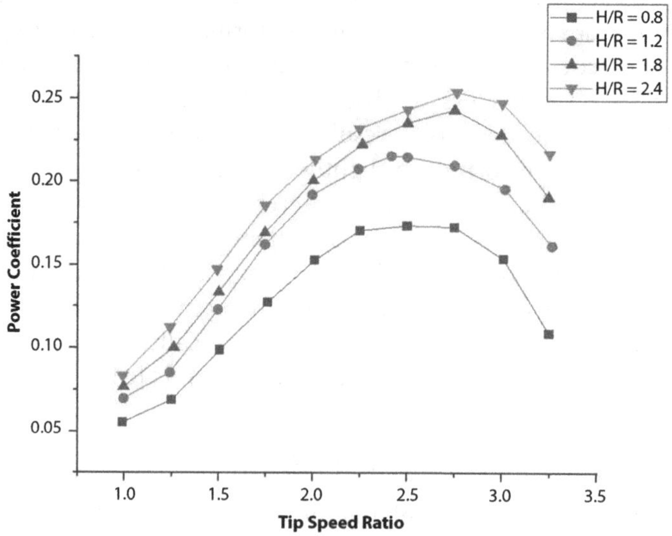

Figure 9.13 Variation of turbine performance for different H/R [63].

have minimum H/R ratio to improve the blade chordal Reynolds number. By raising the radius, the blade chord of VAWT expands, eventually leading to a rise in the chordal Reynolds number.

A study was conducted on a three-dimensional panel model to compare the turbine performance with different H/R, as demonstrated in Figure 9.13 [63]. A comparison of power coefficients was presented for H/R ranging from 0.8 to 2.4. The findings by Li *et al.* illustrated that raising the H/R enhanced the performance, and the best performance was observed for H/R of 2.4 [63].

9.5.3.3 *Blade Aspect Ratio (H/c)*

H/c signifies the relationship between the chord length and blade span [64]. The aspect ratio has a noteworthy impact on the expenses, rigidity, and aerodynamic effectiveness of the system. A range of operational factors and distinct design specifications can influence the aspect ratio of a wind turbine. High aspect ratios are sometimes seen on blades with small diameters operating in light winds. In contrast, larger-sized blades working in high-wind situations show lower aspect ratios [60].

The blade will experience three-dimensional aerodynamics when wind flows over it, allowing wind to distract around the blade tip from high to low pressure area. The generation of vortices at the blade tips due to the circulation phenomenon results in the dissipation of energy and subsequent

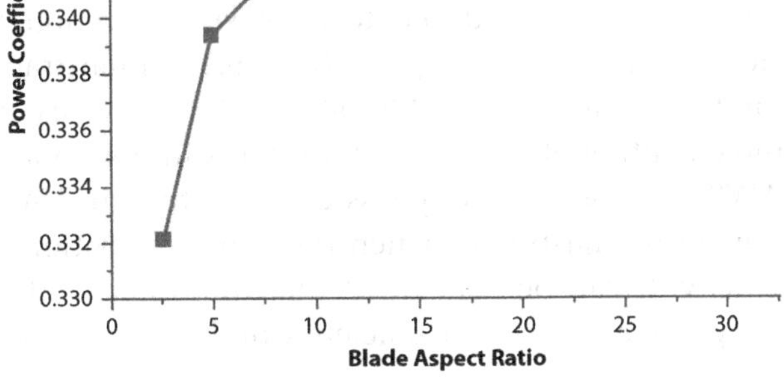

Figure 9.14 Turbine performance for 2-D and 3-D effects for different H/c ratios [57].

production of drag [65]. The aerodynamic properties of a blade with limited length are determined using the Prandtl lifting line theory based on the data obtained from 2D aerodynamic airfoils. The highest power coefficient of the rotor is influenced by the blade aspect ratio, as illustrated in Figure 9.14. The rotor performs better as it asymptotically gets closer to the 2D effect (i.e., infinite blade length) with the increase in blade aspect ratio. The ideal aspect ratio is within the value of 10 to 20, as beyond that the power coefficient does not increase satisfactorily [57].

9.5.4 Pitch

There exist two types of propellers, i.e., fixed and variable pitch [66]. The former has a blade angle that remains constant during operation, while the latter can modify its blade angle to adjust its performance during operation. Fixed-pitch propellers are commonly employed in small aircrafts owing to their cost-effectiveness and simple design. A comparatively higher efficiency level can be achieved by utilizing the variable-pitch mechanism during turbine operation. This mechanism allows the turbine blades to optimize the ratio of blade lift to drag and experience better AOA during their operation. Variable pitch can be present in both active and passive systems. Active systems can induce a change in blade pitch through methods that do not involve any aerodynamic load on the blade [67].

In the context of VAWT, it has been observed that the variable-pitch systems demonstrate better performance in comparison to the fixed-pitch systems. Figure 9.15 illustrates a comparison between fixed and variable-pitch systems, showcasing that the variable-pitch system achieves a higher power coefficient compared to the fixed-pitch system. Introducing variable-pitch systems results in the loss of simplicity and dependability characteristics of the fixed-pitch VAWT technology. This is due to the increased complexity of managing variable-pitch systems, which require more frequent maintenance. The complexity is associated with the maintenance of the sensors employed to detect the wind direction, which may often seek thorough maintenance [64]. Table 9.2 demonstrates the relative benefits and drawbacks of variable-pitch systems over fixed-pitch systems [64]. Implementing blade pitch adjustments in horizontal axis wind turbines (HAWTs) is comparatively less complex than in VAWTs, as the former only requires adjustments when there are fluctuations in the relative flow velocity. It can be concluded that the implementation of the variable-pitch system enhances the efficiency of the turbine [68].

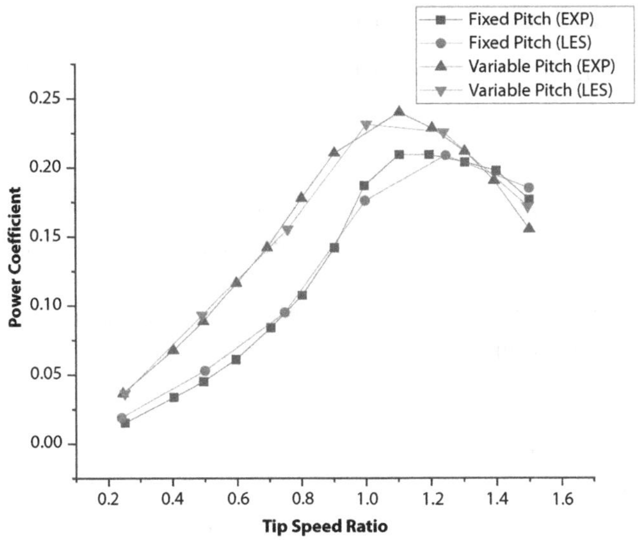

Figure 9.15 Turbine performance for fixed and variable-pitch systems [68].

Table 9.2 Advantages and disadvantages of different pitch systems [69].

Characteristic	Pitch	
	Fixed	**Variable**
Peak efficiency	Less	High
Power coefficient	Better	Poorer
Starting torque	Less	High
Simplicity	More	Less
Reliability	More	Less
Robustness	More	Less
Maintainability	More	Less
Manufacturability and installation	More	Less
System cost	Less	High

9.5.5 Strut Connection Point

The strut connection point refers to a site where the blade of the turbine is attached with the supporting strut. The durability and operational efficiency of the turbine are dependent on its capacity to withstand the mechanical strains induced by the wind and rotational forces. Therefore, the connection point is typically engineered to establish a robust and enduring connection capable of withstanding the applied loads and pressures [69].

The blade pitch angle is created at the blade strut connection (BSC) point by the chord of the blade and the pitch circle's tangent line. When the blade pitch is set to zero, the point of intersection between the tangent line and turbine blade chords is considered neutral. The pitch angles of fixed blades can be categorized as either negative (toe out) or positive (toe in) as demonstrated in Figure 9.16 [69]. The effect of pitch angle on the power coefficient is depicted in Figure 9.17. As observed, 2° pitch angle resulted in the best turbine performance. Moreover, it is observed that an increase in positive pitch angles resulted in reduced overall efficacy. The higher pitch angles (6° or above) can have detrimental effects on turbine performance and can significantly disrupt the balance of blade loading. A study revealed that a high-solidity VAWT with a solidity value of 0.96 experienced a decrease in performance with positive pitch angles [70]. Utilizing a negative blade pitch offers advantages by mitigating the risk of excessively high AOA during the upwind phase. It also increases the chance of too low of an angle of attack during the downhill phase. Moreover, a blade toe-out pitch angle ranging from roughly 1.5° to 3° can accelerate the attainment of the optimal TSR during turbine operation, thereby reducing the turbine's start-up duration by 30% [71].

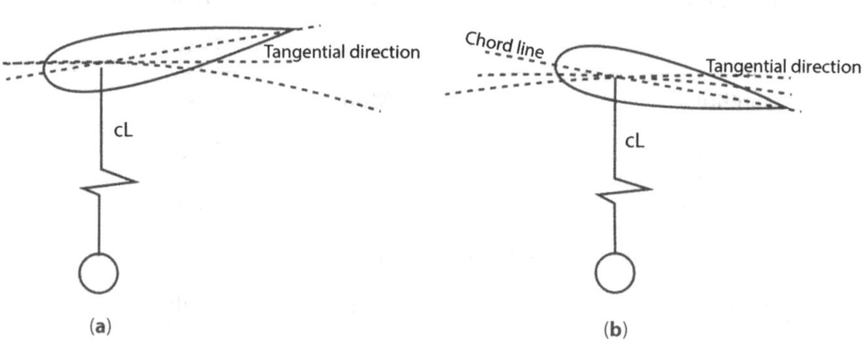

Figure 9.16 Pitch angles for blades: (a) positive and (b) negative.

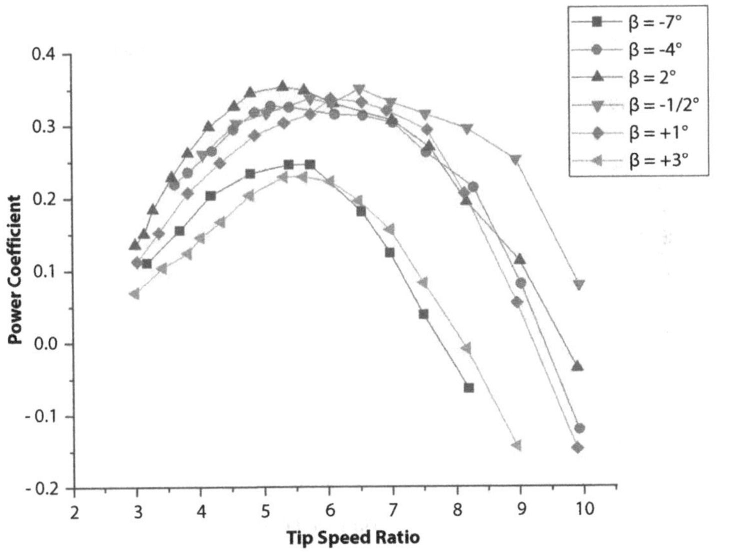

Figure 9.17 Effect of variable-pitch angle on wind turbine performance [70].

9.5.6 Blade Reynolds Number (Re)

The Reynolds number is a dimensionless parameter utilized in fluid dynamics to characterize fluid flow, such as a gas or liquid, around a solid object. Laminar flow is observed at low Reynolds numbers, leading to high lift coefficients and superior aerodynamic efficiency [72]. As a result, there is a rise in the power output and a decrease in drag. Moreover, elevated Reynolds numbers have the potential to induce amplified fluctuating loads on blades, thereby impacting their structural integrity and fatigue life. To achieve optimal wind turbine performance and guarantee safe operation, designers must consider various factors, including the reduction of Reynolds number [73].

It has been observed that the optimal TSR exhibited a negative correlation with the Reynolds number. The phenomenon occurs due to the delay in stall caused by the presence of a turbulent boundary layer that exists on the minimum pressure side of the blade. Figure 9.18 represents the consequence of Reynolds number on the performance of VAWT. Lohry *et al.* observed that in turbulent flow conditions, the VAWT achieves satisfactory performance at specific TSR irrespective of the specific value of Re. According to the research of Lohry *et al.*, upon increasing the Reynolds number, the power coefficient tends to approach a peak value. This implies that a wind energy facility utilizing large-scale VAWTs exhibits greater efficiency than a facility comprising small-scale turbines occupying an

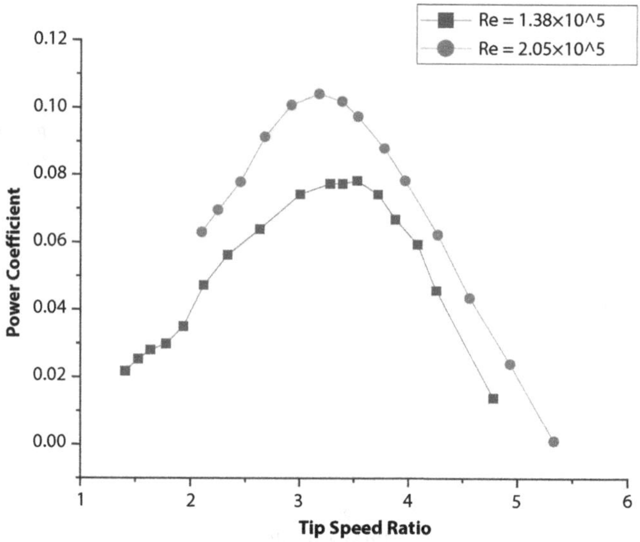

Figure 9.18 Effect of Reynolds number on wind turbine performance [74].

equivalent land area. The self-starting capabilities of VAWTs are positively influenced by a high Reynolds number, thereby promoting the utilization of larger chord lengths in the design [69].

From Figure 9.18, it is apparent that higher Reynolds numbers are generally associated with higher performance coefficients and increasing the TSR improves the power coefficient. In short, an increase in the Reynolds number leads to a larger power coefficient [74].

9.5.7 Strut Effects

The central tower and blades of the VAWT with straight blades are coupled by struts. The struts are essential for delivering torque to the tower, supporting the turbine in strong winds and raising the resonance frequency of the structure itself. Using struts in turbines leads to flow disturbance generating a parasitic or resistive torque, despite its structural advantages [75]. Struts cause the VAWT system to encounter intrinsic drags including profile drag caused by themselves at the junction where the struts join with the blades [76]. Figure 9.19 illustrates the influence of struts on the aerodynamic performance, highlighting that the addition of a support structure has a considerable influence on the turbine performance, particularly at high TSR [77].

In addition, strut thickness also affect the performance of VAWT. Figure 9.20 illustrates the influence of strut thickness on turbine performance.

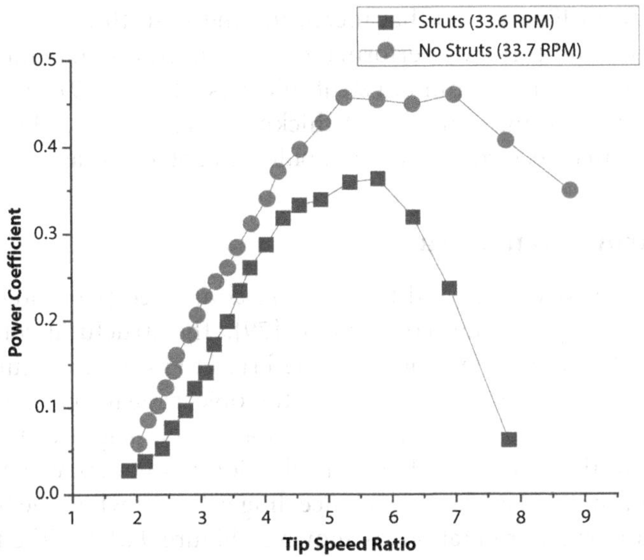

Figure 9.19 The impact of struts on a turbine's performance [78].

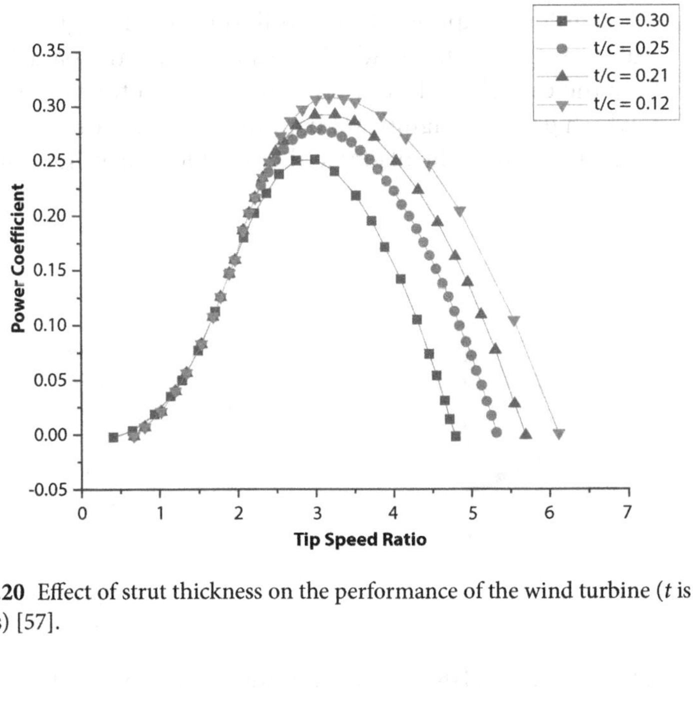

Figure 9.20 Effect of strut thickness on the performance of the wind turbine (t is the strut thickness) [57].

It is clear from Figure 9.20 that increasing the strut thickness negatively impacts the wind turbine's performance. Moreover, the maximum power coefficient was for the minimum strut thickness (t/c = 0.12), and the minimum was for the maximum strut thickness (t/c = 0.30). Therefore, for better performance, strut thickness should be kept as low as possible [57].

9.5.8 Strut Arrangement

The horizontal struts of wind turbines are connected perpendicularly to the blades through T-joint connections [79]. The structural components depicted in Figure 9.21 are commonly referred to as struts. Figure 9.21(a) represents the angular displacement of the object regarding the horizontal plane. This design results in a reduction of the height of the VAWT. According to the aeronautical design rules, joining bodies at sharp angles is not recommended as the interference drag is lowest when the bodies are joined in a normal orientation, as shown in Figure 9.21(b). The tensile or compressive forces will be transmitted to the struts without causing any bending in the tower–strut connection. Hence, orthogonal attachment of the struts is recommended as demonstrated in Figure 9.21(b) [69].

VAWT blades can be supported in various ways using horizontal struts. As shown in Figure 9.22, the strut arrangements can be simply supported, overhung, or cantilever-supported. A cantilever, also recognized as one horizontal supporting strut per blade, is used to reduce the parasitic drag. In cases where the capacity is lower, vertical-axis wind turbines (VAWTs) are utilized, which generate significant bending moments due to centripetal acceleration. To support the blades, either simple or overhung supports

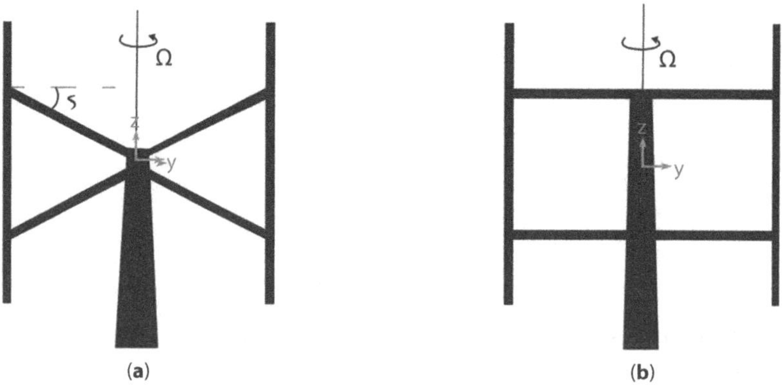

Figure 9.21 (a) Strut angle ς and (b) perpendicular strut connection with blade.

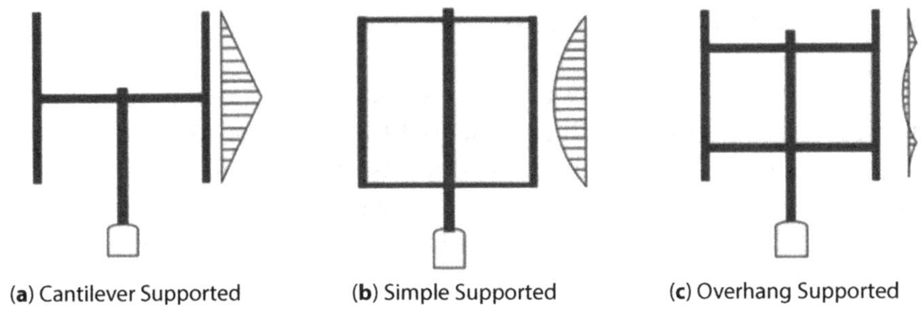

Figure 9.22 Different strut arrangements.

are employed, with the latter utilizing two struts per blade. To lessen the bending moment, it is relevant to position the two struts along the blade in a precise way. The configuration depicted in Figure 9.22(c) is commonly referred to as an overhang-supported configuration [80].

Figure 9.23 summarizes the recommended values of different aerodynamic parameters for efficient wind turbine performance. It is suggested to have less solidity, lower blade numbers, and a fixed-pitch system. All the above-explained parameters must be considered to achieve maximum efficiency of the turbine.

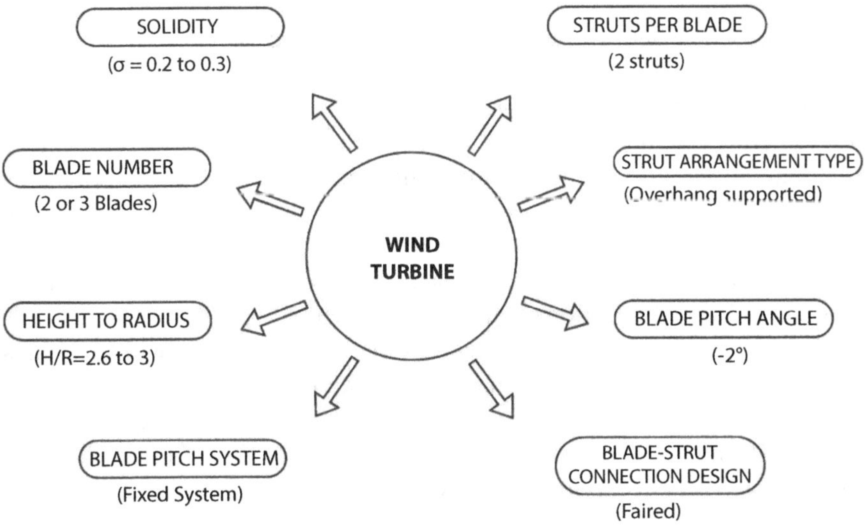

Figure 9.23 Recommended value for the maximum wind turbine performance.

9.6 Wind Turbine Loads

The turbine lifespan depends upon the aerodynamic, gravitational, inertia, and operational loads which are essential to analyze as parts of modeling the wind turbine [81]. The negative impacts of structural loads, especially those brought on by aerodynamic and gravitational forces, become worse as wind turbines get bigger and have higher output power ratings. The output and lifespan of a VAWT can be changed due to the interaction of the structural stresses which leads to an unsatisfactory system performance or system failure. Asymmetric aerodynamic loads caused by variable wind profiles are the main reasons for structural stresses in wind turbines. The overall action of rotor rotation, nacelle yawing, and blade pitching lead to the gyroscopic effect, which is another source of generated structural stresses. This creates cyclic stress on the hub and potentially leads to blade breakdown [82]. The most frequent cause of fatigue stress in large wind turbines is cyclic loads, which, if not addressed, can result in premature turbine failure. Another significant factor contributing to unequal loads across the rotor is the vertical wind shear [82].

Based on the wind velocity, blade speed, AOA, yaw, the lifts, and drags on the blade generate the load on the turbine. The direction of attack is determined by the blade's pitch along with the twist. In order for the generator to move along the axis of rotation, the lift and drag caused by aerodynamics are turned into useful thrusts. The blade must be able to withstand strong pushback forces with little deformation. The widely known BEM theory is used to determine the blade aerodynamic forces [35].

The rotor is a heavy rotating component of the machine that uses an aerodynamic lift to produce torque in the wind, resulting in a significant aerodynamic loading zone [83]. In a rotating system, local aerodynamics are challenging, and cyclic lift augmentation and stall lag can both play key roles [84]. Blade loads can fluctuate significantly because of unsteady inflow and the effects of dynamic stall [83]. Wind shear causes the loads to change once in each rotation of each blade, causing more stresses than those encountered by a downward-pointed blade.

The performance, cost, and physical load endurance of wind turbines are greatly influenced by the material choice made for each component that makes up the wind turbines. Wind turbines can operate in different conditions based on the locations [82]. One of the wind turbine system's most costly components is the rotor, which is fabricated of blades and hubs, and costs around 20% of the price of the wind turbine. Because of the intermittent shifting of aerodynamic stresses and various weather

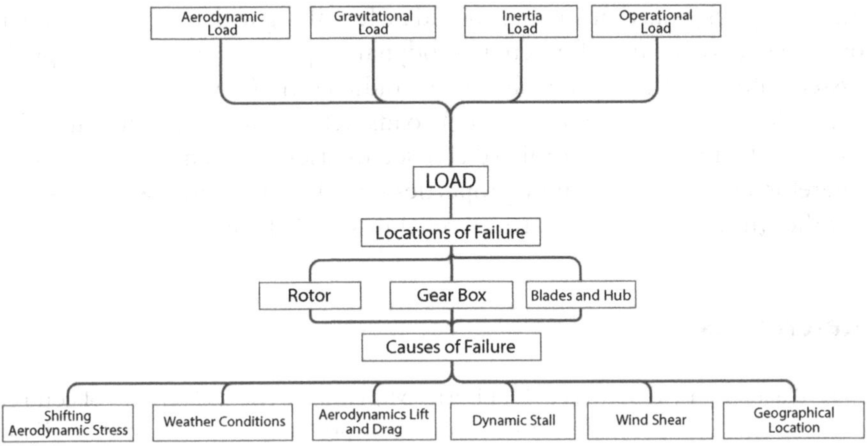

Figure 9.24 Wind turbine loads and causes of failures.

conditions, the blades are more prone to damage. From the rotor to the generator, mechanical power is transferred through gearboxes. Different types of mechanical forces can lead gearboxes to failure. Durability and associated costs are the primary determinants of material choice for wind turbines, especially in offshore applications [82].

In short, the loads on wind turbines and the causes of failures are summarized in Figure 9.24.

9.7 Conclusions

Wind energy extraction is one of the most promising renewable-based approaches to meet the increasing global energy demands. Extensive research has been carried out in recent years to maximize the turbine performance. This chapter illustrated the effects of aerodynamics on turbine performance. It has been observed that different air properties like density, pressure, viscosity, temperature, and humidity directly affect the wind turbine performance. High density and pressure are suitable for better turbine performance as concluded by a number of researchers. However, the performance of wind turbines deteriorates with the increase in air temperature, humidity, and viscosity. This chapter briefly discussed the classical BEM theory for a better understanding of the relationships between aerodynamics variables and turbine performance. Different aerodynamic properties such as solidity, blade numbers, c/R ratio, H/R ratio, Re, pitch angle, and strut effect have significant effects on turbine performance.

This study suggests that solidity should be less along with a smaller number of blades (two or three) for better aerodynamic performance. A fixed-pitch system, along with a suitable height-to-radius ratio (2.6 to 3), is suggested when designing a system. Different loads acting on the turbine and the causes of turbine failure are also discussed in brief at the end of the chapter. Therefore, these aerodynamic properties must be considered when designing the wind turbine for improved reliability and efficiency.

References

1. Blaabjerg, F., Teodorescu, R., Liserre, M., Timbus, A. V., Overview of control and grid synchronization for distributed power generation systems. *IEEE Trans. Ind. Electron.*, 53, 1398, 2006.
2. Sinsel, S.R., Riemke, R.L., Hoffmann, V.H., Challenges and solution technologies for the integration of variable renewable energy sources—a review. *Renew. Energy*, 145, 2271, 2020.
3. Global Wind Energy Council launched. *Refocus*, 6, 11, 2005.
4. Bošnjaković, M., Katinić, M., Santa, R., Marić, D., Wind Turbine Technology Trends. *Appl. Sci.*, 12, 8653, 2022.
5. A. Jamal, S. Suripto, and R. Syahputra, "Multi-band power system stabilizer model for power flow optimization in order to improve power system stability.," *J. Theor. Appl. Inf. Technol.*, vol. 81, no. 1, 2015.
6. A. Betz and Z. Gesamte, "The maximum theoretically achievable utilization of wind power by a wind turbine.," *Turbine Wesen*, vol. 17, p. 320, 1920.
7. N. E. Joukowsky, "Windmill of the NEJ type," *Transactions of the Central Institute for Aero-Hydrodynamics of Moscow*, vol. 1, p. 57, 1920.
8. Glauert, H., Airplane Propellers. In: Durand, W.F. (ed.) *Aerodynamic Theory*. pp. 169–360. Springer Berlin Heidelberg, Berlin, Heidelberg, 1935.
9. Sørensen, J.N., The general momentum theory. In: Sørensen, J.N. (ed.) *Research Topics in Wind Energy*. pp. 43–58. Springer International Publishing, Cham, 2016.
10. Tian, W., Ni, X., Mao, Z., Wang, Y.F., Study on the performance of a new VAWT with overlapped side-by-side Savonius rotors. *Energy Convers. Manag.*, 264, 115746, 2022.
11. Kaya, A.F., Acır, A., Enhancing the aerodynamic performance of a Savonius wind turbine using Taguchi optimization method. *Energy Sources, Part A Recover. Util. Environ. Eff.*, 44, 5610, 2022.
12. Porté-Agel, F., Bastankhah, M., Shamsoddin, S., Wind-Turbine and Wind-Farm Flows: A Review. *Boundary-Layer Meteorol.*, 174, 1, 2020.
13. Rocha, P.A.C., Carneiro de Araujo, J.W., Lima, R.J.P., Vieira da Silva, M.E., Albiero, D., de Andrade, C.F., Carneiro, F.O.M., The effects of blade pitch

angle on the performance of small-scale wind turbine in urban environments. *Energy*, 148, 169, 2018.
14. Lackner, M.A., An investigation of variable power collective pitch control for load mitigation of floating offshore wind turbines. *Wind Energy*, 16, 435, 2013.
15. Castellani, F., Eltayesh, A., Becchetti, M., Segalini, A., Aerodynamic analysis of a wind-turbine rotor affected by pitch unbalance. *Energies*, 14, 745, 2021.
16. Madsen, M.H.A., Zahle, F., Horcas, S.G., Barlas, T.K., Sørensen, N.N., CFD-based curved tip shape design for wind turbine blades. *Wind Energy Sci.*, 7, 1471, 2022.
17. Du, Q., Li, Y., Yang, L., Liu, T., Zhang, D., Xie, Y., Performance prediction and design optimization of turbine blade profile with deep learning method. *Energy*, 254, 124351, 2022.
18. M. Chetan, M. S. Sakib, D. T. Griffith, and S. Yao, "Aero-structural design study of extreme-scale segmented ultralight morphing rotor blades," in *AIAA Aviation 2019 Forum*, 2019, p. 3347.
19. Duthé, G., Abdallah, I., Barber, S., Chatzi, E., Modeling and monitoring erosion of the leading edge of wind turbine blades. *Energies*, 14, 7262, 2021.
20. M. O. L. Hansen, *Aerodynamics of wind turbines*. Routledge, 2015.
21. Adaramola, M.S., Krogstad, P.Å., Experimental investigation of wake effects on wind turbine performance. *Renew. Energy*, 36, 2078, 2011.
22. Shapiro, C.R., Gayme, D.F., Meneveau, C., Modelling yawed wind turbine wakes: A lifting line approach. *J. Fluid Mech.*, 841, R11, 2018.
23. Logan, E., Wind turbines. Cambridge University Press 1981.
24. Sørensen, J.N., Aerodynamic aspects of wind energy conversion. *Annu. Rev. Fluid Mech.*, 43, 427, 2011.
25. Sefidgar, Z., Ahmadi Joneidi, A., Arabkoohsar, A., A Comprehensive Review on Development and Applications of Cross-Flow Wind Turbines. *Sustain.*, 15, 4679, 2023.
26. Song, K., Huan, H., Kang, Y., Aerodynamic Performance and Wake Characteristics Analysis of Archimedes Spiral Wind Turbine Rotors with Different Blade Angle. *Energies*, 16, 385, 2023.
27. Mazzeo, F., Micheletto, D., Talamelli, A., Segalini, A., An Experimental Study on a Wind Turbine Rotor Affected by Pitch Imbalance. *Energies*, 15, 8665, 2022.
28. Wang, X., Tang, Z., Yan, N., Zhu, G., Effect of Different Types of Erosion on the Aerodynamic Performance of Wind Turbine Airfoils. *Sustain.*, 14, 12344, 2022.
29. Wenehenubun, F., Saputra, A., Sutanto, H., An experimental study on the performance of Savonius wind turbines related with the number of blades. *Energy Procedia*, 68, 297, 2015.
30. Johansen, J., Madsen, H.A., Gaunaa, M., Bak, C., Srensen, N.N., Design of a wind turbine rotor for maximum aerodynamic efficiency. *Wind Energy*, 12, 261, 2009.

31. Capon, C.J., Smith, B., Brown, M., Abay, R., Boyce, R.R., Effect of ionospheric drag on atmospheric density estimation and orbit prediction. *Adv. Sp. Res.*, 63, 2495, 2019.
32. Wick, G.A., Emery, W.J., Kantha, L.H., Schlüssel, P., The behavior of the bulk - Skin sea surface temperature difference under varying wind speed and heat flux. *J. Phys. Oceanogr.*, 26, 1969, 1996.
33. Karthikeyan, N., Kalidasa Murugavel, K., Arun Kumar, S., Rajakumar, S., Review of aerodynamic developments on small horizontal axis wind turbine blade. *Renew. Sustain. Energy Rev.*, 42, 801, 2015.
34. Calva, P.A., Espino, F.C., Effect of the humidity in the ionic mobility in reduced air-density. In: *Conference on Electrical Insulation and Dielectric Phenomena (CEIDP)*, Annual Report. pp. 508–511, 1998.
35. Mouhsine, S. El, Oukassou, K., Ichenial, M.M., Kharbouch, B., Hajraoui, A., Aerodynamics and structural analysis of wind turbine blade. *Procedia Manuf.*, 22, 747, 2018.
36. Balat, M., A review of modern wind turbine technology. *Energy Sources, Part A Recover. Util. Environ. Eff.*, 31, 1561, 2009.
37. Lipian, M., Czapski, P., Obidowski, D., Fluid-structure interaction numerical analysis of a small, urban wind turbine blade. *Energies*, 13, 1832, 2020.
38. Li, Q., Maeda, T., Kamada, Y., Murata, J., Shimizu, K., Ogasawara, T., Nakai, A., Kasuya, T., Effect of solidity on aerodynamic forces around straight-bladed vertical axis wind turbine by wind tunnel experiments (depending on number of blades). *Renew. Energy*, 96, 928, 2016.
39. Sørensen, J.N., Aerodynamic aspects of wind energy conversion. *Annu. Rev. Fluid Mech.*, 43, 427, 2011.
40. Bertagnolio, F., Madsen, H.A., Bak, C., Troldborg, N., Fischer, A., Aerodynamic noise characterization of a full-scale wind turbine through high-frequency surface pressure measurements. *Int. J. Aeroacoustics*, 14, 729, 2015.
41. Ahmadi-Baloutaki, M., Carriveau, R., Ting, D.S.K., A wind tunnel study on the aerodynamic interaction of vertical axis wind turbines in array configurations. *Renew. Energy*, 96, 904, 2016.
42. Howell, R., Qin, N., Edwards, J., Durrani, N., Wind tunnel and numerical study of a small vertical axis wind turbine. *Renew. Energy*, 35, 412, 2010.
43. Thiébaut, M., Filipot, J.F., Maisondieu, C., Damblans, G., Jochum, C., Kilcher, L.F., Guillou, S., Characterization of the vertical evolution of the three-dimensional turbulence for fatigue design of tidal turbines: 3D turbulence for fatigue design of TEC. *Philos. Trans. R. Soc. A Math. Phys. Eng. Sci.*, 378, 20190495, 2020.
44. Bottasso, C.L., Campagnolo, F., Petrović, V., Wind tunnel testing of scaled wind turbine models: Beyond aerodynamics. *J. Wind Eng. Ind. Aerodyn.*, 127, 11, 2014.
45. Jung, S.N., No, T.S., Ryu, K.W., Aerodynamic performance prediction of a 30 kW counter-rotating wind turbine system. *Renew. Energy*, 30, 631, 2005.

46. Li, Q., Maeda, T., Kamada, Y., Murata, J., Yamamoto, M., Ogasawara, T., Shimizu, K., Kogaki, T., Study on power performance for straight-bladed vertical axis wind turbine by field and wind tunnel test. *Renew. Energy*, 90, 291, 2016.
47. Bastankhah, M., Porté-Agel, F., A new miniaturewind turbine for wind tunnel experiments. Part I: Design and performance. *Energies*, 10, 908, 2017.
48. Talavera, M., Shu, F., Experimental study of turbulence intensity influence on wind turbine performance and wake recovery in a low-speed wind tunnel. *Renew. Energy*, 109, 363, 2017.
49. Hand, B., Kelly, G., Cashman, A., Aerodynamic design and performance parameters of a lift-type vertical axis wind turbine: A comprehensive review. *Renew. Sustain. Energy Rev.*, 139, 110699, 2021.
50. Brusca, S., Lanzafame, R., Messina, M., Design of a vertical-axis wind turbine: how the aspect ratio affects the turbine's performance. *Int. J. Energy Environ. Eng.*, 5, 333, 2014.
51. McLaren, K., Tullis, S., Ziada, S., Computational fluid dynamics simulation of the aerodynamics of a high solidity, small-scale vertical axis wind turbine. *Wind Energy*. 15, 349, 2012.
52. Guillaud, N., Balarac, G., Goncalvès, E., Zanette, J., Large Eddy Simulations on Vertical Axis Hydrokinetic Turbines - Power coefficient analysis for various solidities. *Renew. Energy*, 147, 473, 2020.
53. Daróczy, L., Janiga, G., Petrasch, K., Webner, M., Thévenin, D., Comparative analysis of turbulence models for the aerodynamic simulation of H-Darrieus rotors. *Energy*, 90, 680, 2015.
54. Sandia National Laboratories, Cost Study for Large Wind Turbine Blades. Office of Scientific and Technical Information (OSTI) 2003.
55. Blackwell, B.F., Sullivan, W.N., Reuter, R.C., Banas, J.F., Engineering development status of the Darrieus wind turbine. *J. Energy*, 1, 50, 1977.
56. Hand, B., Kelly, G., Cashman, A., Numerical simulation of a vertical axis wind turbine airfoil experiencing dynamic stall at high Reynolds numbers. *Comput. Fluids*, 149, 12, 2017.
57. Hand, B., Cashman, A., Conceptual design of a large-scale floating offshore vertical axis wind turbine. *Energy Procedia*, 142, 83, 2017.
58. Rezaeiha, A., Montazeri, H., Blocken, B., Towards optimal aerodynamic design of vertical axis wind turbines: Impact of solidity and number of blades. *Energy*, 165, 1129, 2018.
59. Singh, R.K., Ahmed, M.R., Blade design and performance testing of a small wind turbine rotor for low wind speed applications. *Renew. Energy*, 50, 812, 2013.
60. Liang, Y. Bin, Zhang, L.X., Li, E.X., Liu, X.H., Yang, Y., Design considerations of rotor configuration for straight-bladed vertical axis wind turbines. *Adv. Mech. Eng.*, 2014, 534906, 2014.
61. Van Den Berg, G.P., Wind turbine power and sound in relation to atmospheric stability. *Wind Energy*, 11, 151, 2008.

62. Li, Q., Maeda, T., Kamada, Y., Shimizu, K., Ogasawara, T., Nakai, A., Kasuya, T., Effect of rotor aspect ratio and solidity on a straight-bladed vertical axis wind turbine in three-dimensional analysis by the panel method. *Energy*, 121, 1, 2017.
63. Hand, B., Kelly, G., Cashman, A., Aerodynamic design and performance parameters of a lift-type vertical axis wind turbine: A comprehensive review. *Renew. Sustain. Energy Rev.*, 139, 110699, 2021.
64. Amet, E., Maître, T., Pellone, C., Achard, J.L., 2D numerical simulations of blade-vortex interaction in a darrieus turbine. *J. Fluids Eng. Trans. ASME*, 131, 1111031, 2009.
65. Paraschivoiu, I., Trifu, O., Saeed, F., H-Darrieus wind turbine with blade pitch control. *Int. J. Rotating Mach.*, 2009, 1, 2009.
66. Yang, X., Tang, H., Chen, M., Performance modeling and optimization assessment of variable pitch fan for ultrafan engine. In: *2018 Joint Propulsion Conference*. American Institute of Aeronautics and Astronautics, 2018.
67. Elkhoury, M., Kiwata, T., Aoun, E., Experimental and numerical investigation of a three-dimensional vertical-axis wind turbine with variable-pitch. *J. Wind Eng. Ind. Aerodyn.*, 139, 111, 2015.
68. Goude, A., Bülow, F., Robust VAWT control system evaluation by coupled aerodynamic and electrical simulations. *Renew. Energy*, 59, 193, 2013.
69. Mulleners, K., Fouest, S. Le, Optimal blade pitch control for enhanced vertical-axis wind turbine performance. *Res. Sq.*, 2023.
70. Asr, M.T., Nezhad, E.Z., Mustapha, F., Wiriadidjaja, S., Study on start-up characteristics of H-Darrieus vertical axis wind turbines comprising NACA 4-digit series blade airfoils. *Energy*, 112, 528, 2016.
71. Rice, M., Hall, J., Papadakis, G., Yianneskis, M., Investigation of laminar flow in a stirred vessel at low Reynolds numbers. *Chem. Eng. Sci.*, 61, 2762, 2006.
72. Karthikeyan, N., Kalidasa Murugavel, K., Arun Kumar, S., Rajakumar, S., Review of aerodynamic developments on small horizontal axis wind turbine blade. *Renew. Sustain. Energy Rev.*, 42, 801, 2015.
73. Battisti, L., Benini, E., Brighenti, A., Castelli, M.R., Dell'Anna, S., Dossena, V., Persico, G., Paulsen, U.S., Pedersen, T.F., Normalized performance and load data for the deepwind demonstrator in controlled conditions. *Data Br.*, 8, 1120, 2016.
74. Hand, B., Cashman, A., Aerodynamic modeling methods for a large-scale vertical axis wind turbine: A comparative study. *Renew. Energy*, 129, 12, 2018.
75. Sutherland, H.J., Berg, D.E., Ashwill, T.D., A Retrospective of VAWT Technology. Office of Scientific and Technical Information (OSTI) 2012.
76. Worstell, M.H., Aerodynamic Performance of the Doe/Sandia 17-M-Diameter Vertical-Axis Wind Turbine. *J. Energy*, 5, 39, 1981.
77. Bachant, P., Wosnik, M., Gunawan, B., Neary, V.S., Experimental study of a reference model vertical-axis cross-flow turbine. *PLoS One*, 11, e0163799, 2016.

78. Islam, M., Fartaj, A., Carriveau, R., Analysis of the design parameters related to a fixed-pitch straight-bladed vertical axis wind turbine. *Wind Eng.*, 32, 491, 2008.
79. Ahmadi-Baloutaki, M., Carriveau, R., Ting, D.S.K., Straight-bladed vertical axis wind turbine rotor design guide based on aerodynamic performance and loading analysis. *Proc. Inst. Mech. Eng. Part A J. Power Energy*, 228, 742, 2014.
80. Joselin Herbert, G.M., Iniyan, S., Sreevalsan, E., Rajapandian, S., A review of wind energy technologies. *Renew. Sustain. Energy Rev.*, 11, 1117, 2007.
81. Njiri, J.G., Söffker, D., State-of-the-art in wind turbine control: Trends and challenges. *Renew. Sustain. Energy Rev.*, 60, 377, 2016.
82. Vorpahl, F., Schwarze, H., Fischer, T., Seidel, M., Jonkman, J., Offshore wind turbine environment, loads, simulation, and design. *Wiley Interdiscip. Rev. Energy Environ.*, 2, 548, 2013.
83. Schreck, S.J., Rotationally augmented flow structures and time varying loads on turbine blades. In: *Collection of Technical Papers - 45th AIAA Aerospace Sciences Meeting*. pp. 7584–7596. American Institute of Aeronautics and Astronautics, 2007.
84. Sun, X., Huang, D., Wu, G., The current state of offshore wind energy technology development. *Energy*, 41, 298, 2012.

Index

3d, 137, 161, 174

Adaptive management, 21
Advanced forecasting tools, 46
Aerodynamic challenges, 209
Aerodynamic imbalance, 208
Aerodynamic noise, 36
Aerodynamic stress, 238
Aerodynamically, 140
Aerodynamics, 4, 6, 14, 207, 238, 239
Aeroelasticity, 91
Aesthetic standards, 35
Afforestation, 26
AHP, 188
Air, 88
Air density, 210, 211, 215, 218
Airborne wind energy, 12
Airfoil, 92
Algorithms, 93
Aligning, 90
Alternative energy, 116, 117, 120, 123
American wind energy association, 44
Ampere, 70
Analytic hierarchy, 185
Angle of Attack (AOA), 209, 215, 224, 230, 232, 238
Arena, 136
Artificial intelligence, 91
Aspect ratio, 229, 230

Batteries, 70, 87, 116, 118, 119, 121
Binary matrix, 192
Biodiversity, 137, 163
Biological change, 24
Biological processes, 25
Biomass, 118
Biomass technologies, 34
Blade, 4, 6, 9, 10, 13
Blade element momentum theory, 207, 210, 211, 215
Blade erosion, 209
Blade stiffness, 224
Blade strikes, 27
Blockage ratio, 216
Boundary layer, 208, 211, 233
Broader spectrum, 119
Bucket foundation, 46
Burning, 137

Capacity, 73, 84
Capital, 79
Capture, 134, 135, 160, 161, 167, 168, 169
Captured, 139, 140
Carbon, 2, 10, 14
Carbon fiber, 95
Carbon footprint, 117, 123
Carbon-free emission, 41
Cells, 79
Chemical fuels, 118
Chord, 219, 223, 234
Chord length, 219, 221, 226, 227, 229
Chord/radius ratio, 226
Chronological, 22
Clean energy, 197
Climate, 2

Climate change, 19, 25, 27, 29
Coal, 118, 119
Communication tower, 35
Components, 69
Composite, 133, 149, 150
Compressed air, 118
Compressibility, 74
Computational fluid dynamics (CFD), 209
Computational simulation, 40
Computationally, 24
Conductance, 97
Consistency index, 194
Consumption, 2
Conventional power, 116, 117, 119, 186
Conversion, 5, 6, 88
Converter, 78, 97
Correlation matrix, 197
Cost-effective, 107
Cumulative effect analysis, 25
Curtailment, 99
Cut-in wind speed, 196

Decision matrix, 197
Decommissioned, 135, 136, 147, 151
Deep learning, 209
Delve, 137
Demand, 1, 8, 14
Demographic consequences, 28
Design, 1, 2, 3, 4, 5, 6, 7, 8, 9, 10, 13
Diesel, 120
Direct-drive generators, 54
Disposal, 135-137, 163, 175
Drag, 88
Drag force, 210, 211
Dynamic, 6, 7, 86
Dynamic stall, 213, 224, 227, 238, 239

Eco-friendly, 115, 116
Ecological conformance, 39
Ecological problem, 197
Ecological processes, 21
Economic factors, 224
Ecosystem, 27, 29
Ecosystem health, 29
Efficiency, 1, 4, 8, 9, 10, 13, 14, 88
Efficient wind turbine, 195
Eigenvector, 201
Electre, 187
Electrical grid, 4, 7, 10, 20
Electricity, 1, 12, 63, 84
Electrolysis, 115, 121, 125
Electrolyzer, 72
Emission, 116, 120, 127
Energy, 78, 115, 120, 121, 122, 123, 127, 128
Energy demand, 116, 117, 126
Energy density, 9
Energy Forecasting and Scheduling, 43
Energy production, 34
Energy storage, 4, 10
Energy storage systems, 45
Environment, 1, 9, 12, 18, 116, 122, 123, 125
Environmental considerations, 189
Environmental protection, 25
European renewable energy council, 34

Fabrication, 95
Fatigue, 133, 167, 173, 174, 176
Field testing, 215, 216, 217
Firms, 170, 171
Fixed pitch, 230, 237, 240
Floating foundation, 48
Fluid, 87
Fly wheels, 107, 118
Footprint, 137, 152, 162
Forefront, 134, 135, 150
Fossil fuel, 2, 18, 52, 97
Fossil-based power sources, 197
Frequency, 79, 99
Fuel, 68, 118

Gas, 118, 121, 125
Gas combined cycle, 118
Gasoline, 120

Generator, 5, 8, 10, 13, 65, 115, 117, 127, 196
Genetic algorithm, 188
Geographical scales, 26, 28, 29
Geothermal, 118, 119
German meteorological service, 45
Global benefits, 27
Global wind power, 43
Green economy, 42
Green resources, 115, 116, 119
Greenhouse, 1, 134, 137, 138, 140, 151
Greenhouse gas, 25, 26, 186
Grid, 74

Harmonic, 105
Harness, 138
Harvest rate, 24
Health, 156, 157, 174
Height-to-Radius Ratio, 207, 227, 240
Horizontal axis wind turbine (HAWT), 6, 9, 208, 217, 230
Hybrid, 64
Hydro, 75
Hydrogen, 71
Hydropower, 79
Hydrothermal, 76

Incinerator plant chimneys, 34
Individual effects to population impacts, 23
Induced flow ratio, 239
Induction factor, 212, 213, 215
Integration, 79
Intermittent energy generation, 39
Interorganizational partnership, 38
Isolation, 117

Jacket foundation, 47

Land acquisition, 51
Layout, 101
Legal acceptance, 25
Levelized Cost of Energy, 41
Lift force, 210
Low-carbon, 18
Low-frequency noise, 36
Lpg, 120

Machine, 3, 4, 12
Machine learning, 92
Magnetic field, 7
Magnetic storages, 118
Materials science, 14
Maximum hub height, 196
Mechanical, 76
Melting, 78
Micro hydro, 118
Micro-grid, 65
Microprocessor, 71
Mitigation strategies, 35
Modular, 92
Modular construction, 56
Modulates, 96
Molten salts, 118
Monopile foundation, 47
Monte Carlo methodology, 187

Nacelle, 54, 139, 140
National Institute of Wind Energy, 49
National renewable energy laboratory, 43
Natural world, 26
Necessitating, 138
Net zero emissions, 18
Noise pollution, 14, 20, 27
Normalized value, 197
Number of blades, 220, 223, 224, 240

Ocean, 118
Offshore wind, 17, 19, 20, 25, 26, 27
Oil, 118, 119
Onshore wind installed capacity, 18
Operation, 138, 172, 176
Operational lifespan, 186
Optimized power generation, 57

Pairwise comparison, 197
Peak output, 14

Petroleum, 145, 146, 151, 152
Photocatalysis, 120
Photovoltaic, 77, 118
Pitch, 92, 208, 239
Pitch angle, 207, 239
Placement, 100
Policy, 20, 24, 30
Pollution, 17, 20, 27, 75, 186
Population, 23, 24, 27
Population-level, 24, 29
Potential, 1, 2, 12
Power, 1, 11, 68, 103
Power coefficient, 208, 233, 234, 236
Power density, 196
Power optimization, 209
Predominantly, 135, 136, 141
Public goods, 25, 26

Rated wind speed, 196
Reinforced, 135, 137, 159-161, 163
Reliability, 207, 215, 231, 240
Renewable energy, 1, 14, 25, 28, 38, 63, 197
Renewable portfolio standards, 44
Resin, 143-146, 151, 152, 158, 159
Resources, 2, 10
Reusable, 133, 175-177
Reynolds number, 207, 220, 224, 226, 227, 228, 233, 234
Risk assessment, 27
Risk management, 22
Robustness, 98
Rotational, 138, 140
Rotor, 134, 138, 140, 196
Rotor speed, 219

Safety, 13
Scalability, 133, 158-171, 176
Sensors, 139
Simulation, 158, 170, 176
Social advantages, 23
Social impacts, 52
Solar, 69
Solar power, 186
Solidity, 214, 239, 240
Speed, 4, 8, 13, 75
Sporadic wind energy, 56
Stability, 98
Stokes, 88
Storage batteries, 67
Store, 11, 116, 120, 125
Structural, 133, 147, 174
Strut effects, 207, 234
Suboptimal, 102
Sun, 116, 117, 119, 125
Sun's radiation, 186
Super capacitor, 118
Sustainability, 33
Sustainable, 17, 18, 24, 73, 83, 164, 175-177
Sustainable development, 1, 2, 14, 186
Sustainable power sources, 186
Swept, 138, 160, 174
Swept area, 196, 223, 228
Synchronous, 87
System, 115, 122, 123, 127, 128

Technology, 75, 85
Tension leg platforms, 48
Tethys content, 21
The national meteorological information centre, 44
Thermal, 70
Threshold, 74
Tip speed ratio (tsp), 208, 221, 224, 225, 226, 227
Topsis, 187
Tower, 196
Transmission, 14
Transmission infrastructure, 51
Tripod foundation, 48
Turbine, 1, 13, 14, 87
Turbine blades, 187
Turbine design, 53
Turbine efficiency index, 189

Turbine selection, 185
Turbine system, 50
Turbulence, 69, 138, 157, 174, 210, 226, 228
Turbulence intensity, 216

Variable pitch, 230, 233
Vertical axis wind turbine, 3, 6, 11
Vertical axis wind turbine (VAWT), 208, 236
Visual impact, 14
Visual impacts, 20
Voltage, 106, 196
Vortex, 94

Wake generation, 208
Water, 117, 118, 119, 126
Weighted score, 197
White paper, 21, 22, 26
Wildlife, 20, 21, 23, 26, 27, 28
Wind, 115, 126, 127, 128
Wind energy, 1, 8, 17, 23, 24, 83, 186, 208, 215, 217, 233, 239
Wind energy control, 53
Wind energy potential, 42
Wind facility design, 194
Wind farm, 17, 20, 24, 187
Wind power, 18, 19, 20, 29, 64
Wind power generation, 194
Wind shear stress, 209
Wind speed, 210, 215, 224, 228
Wind tunnel testing, 215, 217
Wind turbine, 207, 224, 228, 240
Wind turbine design, 217, 220
Wind turbine loads, 238, 239
Wind turbines, 20
Wind utilization, 37
Windmill, 3, 4, 9, 67
Worker, 13
Working principle, 9, 10, 11
Worldwide, 118
Worldwide wind energy forecast, 41

Also of Interest

Check out these other forthcoming and published titles from Scrivener Publishing

Books by the same editor from Wiley-Scrivener

ADVANCED REDOX FLOW TECHNOLOGIES, Edited by Inamuddin and Tariq Altalhi, ISBN: 9781119904793. This book serves as a comprehensive guide to redox flow technologies, from their basic principles to their applications. As a result, this book provides a thorough review of craftsmanship in the subject, as well as the potential for future advances.

HANDBOOK OF WATER POLLUTION, Edited by Inamuddin, Tariq Altalhi, and Arwa Alrooqi, ISBN: 9781119904809. This handbook covers a wide range of water pollution issues and remedies. Researchers, students, and other governmental and non-governmental groups will all benefit from this informative guide. In detail, this book will include chapters in the areas of water pollution, analysis, and solutions.

ESSENTIAL OILS: Extraction Methods and Applications, Edited by Inamuddin, Tariq Altalhi, and Jorddy Neves Cruz, ISBN: 9781119829355. Written and edited by some of the world's top experts in the field, this exciting new volume provides up-to-date information on essential oils' chemical structure, therapeutic, and biological actions, and defines their functional uses and evaluates the advantages and disadvantages of their use in a variety of sectors.

PROTON EXCHANGE MEMBRANE FUEL CELLS: Electrochemical Methods and Computational Fluid Dynamics, Edited by Inamuddin, Omid Moradi, and Mohd Imran Ahamed, ISBN: 9781119829331. Edited by one of the most well-respected and prolific engineers in the world and his team, this book provides a comprehensive overview of hydrogen production, conversion, and storage, offering the scientific literature a comprehensive coverage of this important fuel.

MATERIALS FOR HYDROGEN PRODUCTION, CONVERSION, AND STORAGE, Edited by Inamuddin, Tariq Altalhi, Sayed Mohammed Adnan, and Mohammed A. Amin, ISBN: 9781119829348. Edited by one of the most well-respected and prolific engineers in the world and his team, this book provides a comprehensive overview of hydrogen production, conversion, and storage, offering the scientific literature a comprehensive coverage of this important fuel.

FUNDAMENTALS OF SOLAR CELL DESIGN, Edited by Inamuddin, Mohd Imran Ahamed, Rajender Boddula, and Mashallah Rezakazemi, ISBN: 9781119724704. Edited by one of the most well-respected and prolific engineers in the world and his team, this book provides a comprehensive overview of solar cells and explores the history of evolution and present scenarios of solar cell design, classification, properties, various semiconductor materials, thin films, wafer-scale, transparent solar cells, and other fundamentals of solar cell design.

Biofuel Cells, Edited by Inamuddin, Mohd Imran Ahamed, Rajender Boddula, and Mashallah Rezakazemi, ISBN: 9781119724698. This book covers the most recent developments and offers a detailed overview of fundamentals, principles, mechanisms, properties, optimizing parameters, analytical characterization tools, various types of biofuel cells, edited by one of the most well-respected and prolific engineers in the world and his team.

Biodiesel Technology and Applications, Edited by Inamuddin, Mohd Imran Ahamed, Rajender Boddula, and Mashallah Rezakazemi, ISBN: 9781119724643. This outstanding new volume provides a comprehensive overview on biodiesel technologies, covering a broad range of topics and practical applications, edited by one of the most well-respected and prolific engineers in the world and his team.

Applied Water Science Volume 1: Fundamentals and Applications, Edited by Inamuddin, Mohd Imran Ahamed, Rajender Boddula and Tauseef Ahmad Rangreez, ISBN: 9781119724766. Edited by one of the most well-respected and prolific engineers in the world and his team, this is the first volume in a two-volume set that is the most thorough, up-to-date, and comprehensive volume on applied water science available today.

Applied Water Science Volume 2: Remediation Technologies, Edited by Inamuddin, Mohd Imran Ahamed, Rajender Boddula and Tauseef Ahmad Rangreez, ISBN: 9781119724735. The second volume in a new two-volume set on applied water science, this book provides understanding, occurrence, identification, toxic effects and control of water pollutants in aquatic environment using green chemistry protocols.

Potassium-Ion Batteries: Materials and Applications, Edited by Inamuddin, Rajender Boddula, and Abdullah M. Asiri, ISBN: 9781119661399. Edited by one of the most well-respected and prolific engineers in the world and his team, this is the most thorough, up-to-date, and comprehensive volume on potassium-ion batteries available today.

Rechargeable Batteries: History, Progress, and Applications, Edited by Rajender Boddula, Inamuddin, Ramyakrishna Pothu, and Abdullah M. Asiri, ISBN: 9781119661191. Edited by one of the most well-respected and prolific engineers in the world and his team, this is the most thorough, up-to-date, and comprehensive volume on rechargeable batteries available today.

Printed in the USA/Agawam, MA
September 16, 2026

910661.066